NUMERICAL ANALYSIS

A Second Course

Computer Science and Applied Mathematics

A SERIES OF MONOGRAPHS AND TEXTBOOKS

Editor
Werner Rheinboldt
University of Maryland

NUMERICAL ANALYSIS

A Second Course

JAMES M. ORTEGA
University of Maryland

ACADEMIC PRESS   New York and London

131682

ACADEMIC PRESS, INC.
111 Fifth Avenue, New York, New York 10003

United Kingdom Edition published by
ACADEMIC PRESS, INC. (LONDON) LTD.
24/28 Oval Road, London NW1

LIBRARY OF CONGRESS CATALOG CARD NUMBER: 75-182669

AMS (MOS) 1970 Subject Classification: 65-01

PRINTED IN THE UNITED STATES OF AMERICA

TO MY MOTHER

CONTENTS

PART IV
ROUNDING ERROR

Chapter 9   Rounding Error for Gaussian Elimination

PREFACE

The primary goal of this work is to present some of the basic theoretical results pertaining to the three major problem areas of numerical analysis: rounding error, discretization error, and convergence error (interpreted here as pertaining only to the convergence of sequences and, consequently, to the convergence of iterative methods). Indeed, the organization of the book is in terms of these basic problems and the usual areas of linear equations, differential equations, etc. play a subsidiary role.

An important and pervading aspect of the analysis of the three basic types of error mentioned above is the general notion of stability, a concept which appears in several different guises. The question of mathematical instability, or "ill conditioning" in the numerical analysis parlance, is treated separately, in Part I, as a backdrop to the basic errors. On the other hand, results concerning the mathematical and numerical stability of particular methods are scattered throughout the rest of the book.

Notes on which this book is based have been used for some time at the University of Maryland for a one-semester first-year graduate course for computer science and mathematics majors. As such, it has provided a common knowledge in the theory and analysis of numerical methods for students who arrive with a wide range of backgrounds, and who may or may not pursue more advanced work in numerical analysis.

This course, however, may also be taught on the undergraduate level. In fact, the only prerequisites are courses in linear algebra and advanced calculus; moreover, certain portions of linear algebra of particular interest to us are reviewed in Chapter 1. It is also tacitly presumed that the student has had some undergraduate course in numerical analysis and is familiar with such common numerical methods as gaussian elimination for linear equations, Newton's method for nonlinear equations, Runge–Kutta and multistep methods for differential equations, etc. However, the material is essentially self-contained and it would be possible for the well-motivated and mature student to take this course without prior work in numerical analysis. Finally, some knowledge of ordinary differential equations would be useful but is not essential.

The author would like to express his thanks to Professors Werner Rheinboldt and James Vandergraft of the University of Maryland, to Professor Willian Gragg of the University of California at San Diego, and, especially, to Professor Jorge Moré of Cornell University for their many helpful comments and suggestions, and to Mrs. Dawn Shifflett for her expert typing of the manuscript.

LIST OF COMMONLY USED SYMBOLS

R^n	Real n-dimensional space
C^n	Complex n-dimensional space
$\mathbf{x}^T$	Transpose of the vector $\mathbf{x}$
$\mathbf{x}^H$	Conjugate transpose of the vector $\mathbf{x}$
$L(R^n, R^m)$	Linear operators from R^n to R^m $[L(R^n)$ if $m = n]$
A^{-1}	Inverse of the matrix A
det A	Determinant of the matrix A
$\text{diag}(a_1, \ldots, a_n)$	Diagonal matrix
$\rho(A)$	Spectral radius of A
$\mathbf{e}_i$	ith coordinate vector $(0, \ldots, 0, 1, 0, \ldots, 0)^T$
$\| \ \|$	An arbitrary norm
$\| \ \|_p$	The l_p norm
$K(A)$	The condition number $\|A\| \, \|A^{-1}\|$
$K_p(A)$	$\|A\|_p \|A^{-1}\|_p$
$R^m \times R^n$	The cartesian product $= \{(\mathbf{x}, \mathbf{y}): \mathbf{x} \in R^m, \mathbf{y} \in R^n\}$
$F: R^n \to R^m$	A mapping with domain in R^n and range in R^m
$o(h)$	$o; f(h) = o(h^p)$ implies that $h^{-p}f(h) \to 0$ as $h \to 0$
$O(h^p)$	$O; f(h) = O(h^p)$ implies that $h^{-p}f(h)$ is bounded as $h \to 0$
$\cup, \cap$	Set union, intersection
∂_i	Partial derivative
$\forall$	For all
\$\$\$	End of proof

INTRODUCTION

Numerical analysis has been defined† as "the study of methods and procedures used to obtain approximate solutions to mathematical problems." While this definition is rather broad, it does pinpoint some of the key issues in numerical analysis, namely, *approximate* solution (there is usually no reasonable hope of obtaining the exact solution); *mathematical* problems; the study of *methods* and *procedures*. Let us illustrate and expand on this definition in the context of a particular example.

Consider a second-order nonlinear differential equation

$$y''(t) = f(y(t)), \qquad a \le t \le b \tag{1}$$

together with the boundary conditions

$$y(a) = \alpha, \qquad y(b) = \beta. \tag{2}$$

Here f is a given function and α and β given constants. We assume that this boundary value problem has a unique solution; but, in general, there will be no "closed form" representation for it, and it is necessary to consider methods which will yield an approximation to the solution. One of the common methods for this problem is the **finite difference method**: First divide the interval $[a, b]$ by **grid points**

$$a = t_0 < t_1 < \cdots < t_{n+1} = b;$$

for simplicity, we will assume that the points t_i are equally spaced with

† S. Parter, *Comm. ACM* **12** (1969), 661–663.

spacing h, that is, $t_i = a + ih$, $i = 0, \ldots, n + 1 = (b - a)/h$. Next, approximate $y''(t_i)$ by the standard second difference quotient:

$$y''(t_i) \doteq \frac{1}{h^2} [y(t_{i+1}) - 2y(t_i) + y(t_{i-1})], \qquad i = 1, \ldots, n.$$

Hence, using this approximation in (1), we have

$$y(t_{i+1}) - 2y(t_i) + y(t_{i+1}) \doteq h^2 f(y(t_i)), \qquad i = 1, \ldots, n$$

and if we solve the system of n (in general, nonlinear) equations

$$y_{i+1} - 2y_i + y_{i-1} = h^2 f(y_i), \qquad i = 1, \ldots, n, \quad y_0 = \alpha, \quad y_{n+1} = \beta \quad (3)$$

the solution component y_i is, one hopes, an approximation to the exact solution $y(t_i)$ of (1) at the point t_i.

In general, it will also be impossible to find an explicit solution of the equations (3), and one will obtain an approximate solution by means of some iterative process; for example, one of the simplest such iterative processes for (3) is

$$y_i^{k+1} = y_i^k - \frac{[2y_i^k - y_{i-1}^k - y_{i+1}^k + h^2 f(y_i^k)]}{2 + h^2 f'(y_i^k)}, \qquad i = 1, \ldots, n \quad (4)$$

where the superscript k denotes the iteration number. (This is sometimes called the **Newton–Jacobi** method; a fuller discussion of iterative methods will be given in Part III.) Now even if the iterates converge to the solution, that is, if

$$\lim_{k \to \infty} y_i^k \to y_i, \qquad i = 1, \ldots, n$$

we will be forced to stop after some finite number K of iterations and take y_i^K as an approximation to y_i. But, in general, we will not be able to compute y_i^K exactly because of the rounding error in carrying out the calculation (4). Hence we will end up with, say, $\hat{y}_i^K$, as an approximation to y_i and hence to $y(t_i)$.

The above example illustrates the three major sources of error in the study of numerical methods. Given any "continuous problem"—for example, any differential or integral equation—the first step toward its approximate solution is the reduction to a "discrete analogue" involving only finitely many variables. Thus, the equations (3) form a discrete analogue for (1) and (2). The variables in the discrete analogue need not be the actual function values of the continuous solution at selected points but may be Fourier coefficients, etc. In any case, the difference between the exact solution of the discrete analogue and the exact solution of the original solution is called the (global) **discretization error**, and a major problem of numerical analysis is to give estimates for this error, usually in terms of the data of the original problem together with parameters of the discretization. For the example above, one would be interested in knowing the behavior of $y_i = y_i(h)$ as $h \to 0$; in particular, does the solution of the discrete problem tend to that of the original and, if so, how fast. In Part II we shall obtain estimates for the discretization error for both initial and boundary value problems for ordinary differential equations.

The second major source of error comes from terminating an infinite sequence at a finite number of terms; this will be called **convergence error**. This error arises in numerous contexts—for example, summing an infinite series—but its most prevalent occurrence is with the use of iterative processes. In the above example, the convergence error is, of course, the difference $|y_i^K - y_i|$. We will investigate this type of error primarily in Part III for certain common iterative methods.

Finally, there is the question of rounding error which pervades essentially every calculation. Chapter 9 will be devoted to obtaining bounds on the rounding error in solving linear equations by elimination.

Underlying all three of the above sources of error is the general notion of **instability**, which appears in various guises and situations but usually in the sense of "small changes produce large changes." First, there is the question of the stability of the solution of the problem to be solved; generally, the solution will be said to be stable if "small" changes in the data of the problem produce only "small" changes in the solution. For example, if small changes in the boundary values α and β of (2) cause only small changes in the solution of (1), then we would say that the solution is stable (with respect to the boundary values) whereas if large changes in the solution occur, then it is unstable or—in the more usual parlance of numerical analysis—**ill conditioned**. The problem of ill conditioning effects essentially every large class of mathematical problems, and in Part I we shall examine several examples of both stable and unstable problems.

Secondly, there is the stability of the discrete analogue. It may be, for example, that the solution of the original problem is stable but that of the discrete analogue is unstable. As we shall see in Chapter 5, this is a severe potential problem for initial value problems particularly. Since the most common discrete analogues of differential equations are difference equations [(3) may be considered as a difference equation], this leads us to consider in Chapter 4 the stability of solutions of difference equations as well as solutions of differential equations.

Since any iterative method may be viewed as a difference equation, and the criteria for (asymptotic) stability for solutions of initial value problems for difference equations correspond exactly to criteria for (local) convergence for iterative processes, hence stability—in this technical sense—is equivalent to certain questions in the study of convergence error for sequences. This connection will be explored more precisely in Part III.

Finally, there is the question of "numerical stability" or stability under rounding error. In the sense used here this will mean that the sequence of arithmetic operations embodying the final computer algorithm does not lead to a catastrophic buildup of rounding error for, at least, well-conditioned problems. For example, the gaussian elimination procedure without row interchanges is a (potentially) unstable method in the above sense. This type of stability is interwoven with the rounding error analysis and will be considered in Chapter 9.

CHAPTER I

LINEAR ALGEBRA

The most important tool in many areas of numerical analysis is linear algebra and matrix theory. This is certainly and quite naturally true for those computational problems that arise in linear algebra: solution of linear systems of equations, computation of eigenvalues and eigenvectors of a matrix, etc. But it is also true for a surprisingly large number of other problem areas: nonlinear equations, differential equations, approximation theory, etc. in which many times the analysis of the corresponding numerical methods hinges crucially on results from linear algebra. In more advanced work, infinite-dimensional linear algebra—functional analysis—plays an analogous role.

In any case, linear algebra will provide much of the background for this book, and we will review in this chapter many of those basic parts of linear algebra which will be of most value to us. Additional, more specialized results are also scattered throughout other sections of the book.

I.I EIGENVALUES AND CANONICAL FORMS

We denote by R^n the real n-dimensional space of column vectors $\mathbf{x}$ with components $x_1, \ldots, x_n$ and by C^n the corresponding complex space. For $\mathbf{x} \in R^n$, $\mathbf{x}^\mathrm{T}$ will denote the **transpose**, which is the row vector $(x_1, \ldots, x_n)$, while if $\mathbf{x} \in C^n$, $\mathbf{x}^\mathrm{H} = (\bar{x}_1, \ldots, \bar{x}_n)$ is the **conjugate transpose**.

The collection of linear operators from R^m to R^n—or equivalently, as the context dictates, the set of real $n \times m$ matrices—will be denoted by $L(R^m, R^n)$ or simply $L(R^n)$ if $m = n$. Similarly, the set of complex $n \times m$

matrices is denoted by $L(C^m, C^n)$ or $L(C^n)$ if $m = n$. The elements of a matrix $A \in L(C^m, C^n)$ will be written as a_{ij}. We note that when we write $A \in L(C^m, C^n)$ we by no means preclude the possibility that A is, in fact, real.

If $A \in L(R^m, R^n)$, then A^T will denote the **transpose** of A, while if A is complex, then A^H will denote the **conjugate transpose**. For any $n \times n$ matrix A, det A is the **determinant** of A, and A^{-1} the **inverse**. We will say that A is **nonsingular** if A^{-1} exists. We recall the following basic theorem on invertibility.

1.1.1   Let $A \in L(C^n)$. Then the following are equivalent:

(a) A is nonsingular;
(b)   det $A \neq 0$;
(c)   the linear system $A\mathbf{x} = 0$ has only the solution $\mathbf{x} = \mathbf{0}$;
(d)   for any vector $\mathbf{b}$, the linear system $A\mathbf{x} = \mathbf{b}$ has a unique solution;
(e)   The columns (and rows) of A are linearly independent; that is, if $\mathbf{u}_1, \ldots, \mathbf{u}_n$ are the columns of A and $\alpha_1 \mathbf{u}_1 + \cdots + \alpha_n \mathbf{u}_n = \mathbf{0}$, then the scalars α_i are all zero.

The last condition (1.1.1e) may be rephrased to say that A has rank n where, in general, the **rank** is defined as the number of linearly independent columns (or rows) of the matrix.

If $A \in L(C^n)$, then a (real or complex) scalar λ and a vector $\mathbf{x} \neq 0$ are an **eigenvalue** and **eigenvector** of A if

$$A\mathbf{x} = \lambda\mathbf{x}. \tag{1}$$

By Theorem 1.1.1 it follows that λ is an eigenvalue if and only if

$$\det(A - \lambda I) = 0; \tag{2}$$

this is the **characteristic equation** of A. (Here, as always, I is the identity matrix.) Consequently A has precisely n (not necessarily distinct) eigenvalues, the n roots of (2). The collection of these n eigenvalues $\lambda_1, \ldots, \lambda_n$ is called the **spectrum** of A and

$$\rho(A) = \max_{1 \leq i \leq n} |\lambda_i| \tag{3}$$

is the **spectral radius** of A.

Eigenvalues are, in general, difficult to compute, but there is an important class of matrices in which they are available by inspection. These are (upper or lower) **triangular** matrices:

$$A = \begin{bmatrix} a_{11} & \cdots & a_{1n} \\ & \ddots & \vdots \\ \bigcirc & & a_{nn} \end{bmatrix}, \qquad A = \begin{bmatrix} a_{11} & & \bigcirc \\ \vdots & \ddots & \\ a_{n1} & \cdots & a_{nn} \end{bmatrix}.$$

Clearly, the eigenvalues of a triangular matrix are just the main diagonal elements. An important special case of triangular matrices are **diagonal** matrices

$$D = \begin{bmatrix} d_1 & & \bigcirc \\ & \ddots & \\ \bigcirc & & d_n \end{bmatrix}$$

which we will usually denote by $D = \operatorname{diag}(d_1, \ldots, d_n)$.

For a general matrix $A \in L(C^n)$, the eigenvalues of A may have an arbitrary distribution; but for certain common and important classes of matrices, the eigenvalues are constrained to lie in certain portions of the complex plane. For example, $A \in L(R^n)$ is **orthogonal** if $A^T = A^{-1}$ and $A \in L(C^n)$ is **unitary** if $A^H = A^{-1}$, and the eigenvalues of either type of matrix are all of magnitude one (E1.1.1). Similarly, $A \in L(R^n)$ is **symmetric** if $A^T = A$, and $A \in L(C^n)$ is **hermitian** if $A^H = A$. The eigenvalues of either a symmetric or hermitian matrix are all real (E1.1.2). Moreover, a matrix $A \in L(R^n)$ is **positive semidefinite** if

$$\mathbf{x}^T A \mathbf{x} \geq 0, \qquad \mathbf{x} \in R^n \tag{4}$$

and **positive definite** if strict inequality holds in (4) when $\mathbf{x} \neq 0$. The analogous definition holds for complex matrices with $\mathbf{x}^T$ replaced by $\mathbf{x}^H$. The eigenvalues of a symmetric or hermitian positive semidefinite matrix A are all nonnegative and are all positive if A is positive definite (E1.1.4).

One of the most important operations in matrix theory is the similarity transformation. Two matrices $A, B \in L(C^n)$ are said to be **similar** if there is a nonsingular $P \in L(C^n)$ such that

$$B = P^{-1}AP. \tag{5}$$

This operation arises naturally in considering a linear change of variables; that is, consider the equation

$$\mathbf{y} = A\mathbf{x} \tag{6}$$

and the change of variables

$$\hat{\mathbf{y}} = P^{-1}\mathbf{y}, \qquad \hat{\mathbf{x}} = P^{-1}\mathbf{x}.$$

Then in the variables $\hat{\mathbf{x}}$, $\hat{\mathbf{y}}$, the equation (6) assumes the form

$$\hat{\mathbf{y}} = B\hat{\mathbf{x}}$$

where B is given by (5).

A basic property of the similar transformation is given by the following.

1.1.2   If $A, B \in L(C^n)$ are similar, then A and B have the same eigenvalues.

Proof: We show that the characteristic polynomials of A and B are identical. This is immediate from the product rule for determinants, which yields

$$\det(A - \lambda I) = \det(P^{-1}P)\det(A - \lambda I)$$
$$= \det P^{-1}\det(A - \lambda I)\det P = \det(P^{-1}AP - \lambda I). \quad \$\$\$$$

An alternative proof for 1.1.2 is that, clearly, $A\mathbf{x} = \lambda\mathbf{x}$ if and only if $P^{-1}AP\mathbf{y} = \lambda\mathbf{y}$ where $\mathbf{y} = P^{-1}\mathbf{x} \neq \mathbf{0}$.

We prove next an important result about symmetric matrices.

1.1.3   Let $A \in L(R^n)$ be symmetric. Then there is a real orthogonal matrix P such that $P^T A P$ is diagonal.

Proof: Let $\lambda_1, \ldots, \lambda_n$ be the eigenvalues of A and $\mathbf{x}_1$ an eigenvector corresponding to λ_1 and with $\mathbf{x}_1^T\mathbf{x}_1 = 1$; since the λ_i are real (E1.1.2), we may assume that $\mathbf{x}_1$ is real. Let $\mathbf{u}_1, \ldots, \mathbf{u}_{n-1}$ be $n - 1$ mutually orthogonal vectors all orthogonal to $\mathbf{x}_1$ and with $\mathbf{u}_i^T\mathbf{u}_i = 1$, $i = 1, \ldots, n - 1$. Next, define the $n \times (n - 1)$ matrix $U_1 = (\mathbf{u}_1, \ldots, \mathbf{u}_{n-1})$ with columns $\mathbf{u}_i$, and the corresponding $n \times n$ matrix $P_1 = (\mathbf{x}_1, U_1)$ with first column equal to $\mathbf{x}_1$. Then the orthogonality conditions imply that P_1 is orthogonal and $U_1^T\mathbf{x}_1 = \mathbf{0}$, so that

$$P_1^T A P_1 = \begin{bmatrix} \mathbf{x}_1^T \\ U_1^T \end{bmatrix}(\lambda_1\mathbf{x}_1, AU_1) = \begin{bmatrix} \lambda_1 & 0 \\ 0 & U_1^T A U_1 \end{bmatrix}.$$

Since

$$\det(A - \lambda I) = \det(P_1^{\mathrm{T}} A P_1 - \lambda I) = (\lambda_1 - \lambda)\det(U_1^{\mathrm{T}} A U_1 - \lambda I)$$

the $(n-1) \times (n-1)$ symmetric matrix $A_2 = U_1^{\mathrm{T}} A U_1$ has eigenvalues $\lambda_2, \ldots, \lambda_n$. Therefore, we may perform the same transformation on A_2 and conclude that there is an $(n-1) \times (n-1)$ orthogonal matrix Q_2 so that

$$Q_2^{\mathrm{T}} A_2 Q_2 = \begin{bmatrix} \lambda_2 & 0 \\ 0 & A_3 \end{bmatrix}$$

where A_3 is an $(n-2) \times (n-2)$ symmetric matrix whose eigenvalues are $\lambda_3, \ldots, \lambda_n$. But then

$$P_2^{\mathrm{T}} P_1^{\mathrm{T}} A P_1 P_2 = \begin{bmatrix} \lambda_1 & & \\ & \lambda_2 & \\ & & A_3 \end{bmatrix}$$

where P_2 is the orthogonal matrix

$$P_2 = \begin{bmatrix} 1 & 0 \\ 0 & Q_2 \end{bmatrix}.$$

Continuing in this way, it is clear that we can construct a sequence of orthogonal matrices $P_1, \ldots, P_n$ so that

$$P_n^{\mathrm{T}} \cdots P_1^{\mathrm{T}} A P_1 \cdots P_n = \operatorname{diag}(\lambda_1, \ldots, \lambda_n)$$

and then the product $P = P_1 \cdots P_n$ is also orthogonal. $\$\$\$$

We note that the same result holds for hermitian matrices if P is taken to be unitary.

Let

$$D = \operatorname{diag}(\lambda_1, \ldots, \lambda_n)$$

be the diagonal matrix of 1.1.3, where the λ_i are the eigenvalues of A. Let $\mathbf{p}_i$, $i = 1, \ldots, n$, denote the columns of P. Then equating the columns of

$$AP = PD \tag{7}$$

yields

$$A\mathbf{p}_i = \lambda_i \mathbf{p}_i, \qquad i = 1, \ldots, n; \tag{8}$$

that is, the columns of P are the eigenvectors of A. Hence, an equivalent form of 1.1.3 is that a real symmetric matrix has n orthogonal eigenvectors.

The geometric interpretation of this result is as follows. Consider the relation

$$\mathbf{x}^T A \mathbf{x} = c = \text{constant} \tag{9}$$

which is the equation of a "conic section" in R^n. For example, if A is positive definite, (9) is the equation of an ellipsoid. Now make the change of variable $\hat{\mathbf{x}} = P^T \mathbf{x}$, where P is as in 1.1.3. Then (9) takes the form

$$c = \hat{\mathbf{x}}^T D \hat{\mathbf{x}} = \sum_{i=1}^{n} \lambda_i \hat{x}_i^2$$

which is the equation of a conic section in standard form. This is known as the **reduction to principal axes** of the equation (9), and the eigenvectors of A give the directions of the principal axes of the conic section.

Theorem 1.1.3 shows that a real symmetric matrix is similar to a diagonal matrix. Is this result true for an arbitrary square matrix? Unfortunately, the answer is no, as will be seen for the simple matrix

$$A = \begin{bmatrix} 0 & 1 \\ 0 & 0 \end{bmatrix}. \tag{10}$$

We first prove the following necessary and sufficient condition.

1.1.4 A matrix $A \in L(C^n)$ is similar to a diagonal matrix if and only if A has n linearly independent eigenvectors.

Proof: If $P^{-1}AP = D$, where D is diagonal, the relations (7) and (8) show that the columns of P are eigenvectors, necessarily linearly independent since P is nonsingular. Conversely, if there are n linearly independent eigenvectors, these may be taken as the columns of P. $\$\$\$

On the basis of this result another important sufficient condition for similarity to a diagonal matrix may be given.

1.1.5 Assume that $A \in L(C^n)$ has n distinct eigenvalues. Then A is similar to a diagonal matrix.

Proof: Let $\lambda_1, \ldots, \lambda_n$ be the eigenvalues of A and $\mathbf{x}_1, \ldots, \mathbf{x}_n$ the corresponding eigenvectors. By 1.1.4, it suffices to prove that the $\mathbf{x}_i$ are linearly independent; that is, if

$$\alpha_1 \mathbf{x}_1 + \cdots + \alpha_n \mathbf{x}_n = 0 \tag{11}$$

then all α_i are zero.

Let j be a fixed but arbitrary index; we will show that $\alpha_j = 0$. Define the matrix

$$U = \prod_{\substack{i=1 \\ i \neq j}}^{n} (A - \lambda_i I)$$

and recall that any two matrices of the form $A - \alpha I$, $A - \beta I$ commute. Hence

$$U \mathbf{x}_k = \left[\prod_{i \neq j, k} (A - \lambda_i I) \right] (A - \lambda_k I) \mathbf{x}_k = 0, \qquad k \neq j$$

so that, by (11),

$$0 = U(\alpha_1 \mathbf{x}_1 + \cdots + \alpha_n \mathbf{x}_n) = \alpha_j U \mathbf{x}_j.$$

But

$$U \mathbf{x}_j = \prod_{i \neq j} (A - \lambda_i I) \mathbf{x}_j = \prod_{i \neq j} (\lambda_j - \lambda_i) \mathbf{x}_j \neq 0$$

since the eigenvalues are distinct. Hence $\alpha_j = 0$. $$$

We return to the matrix of (10). Clearly, this matrix has only one linearly independent eigenvector since in the relation

$$\begin{bmatrix} 0 & 1 \\ 0 & 0 \end{bmatrix} \begin{bmatrix} x_1 \\ x_2 \end{bmatrix} = 0$$

the component x_2 must always be zero. Hence, by 1.1.4, this matrix is not similar to a diagonal matrix.

We now ask: What, in general, is the "simplest" form a matrix can assume under a similarity transformation? We give, without proof, two answers of general usefulness.

1.1.6 (Schur's Theorem) Let $A \in L(C^n)$. Then there is a nonsingular matrix $P \in L(C^n)$ such that $P^{-1}AP$ is triangular. Moreover, P may be chosen to be unitary.

1.1.7   (Jordan Canonical Form Theorem)   Let $A \in L(C^n)$.   Then A is similar to a block diagonal matrix

$$J = \begin{bmatrix} J_1 & & & \\ & J_2 & & \\ & & \ddots & \\ & & & J_m \end{bmatrix}$$

where each J_i is either the 1×1 matrix (λ_i) or a matrix of the form

$$J_i = \begin{bmatrix} \lambda_i & 1 & & \\ & \ddots & \ddots & \\ & & \ddots & 1 \\ & & & \lambda_i \end{bmatrix}$$

where λ_i is an eigenvalue.

We will illustrate the last theorem by various examples and comments. Note first that if A has distinct eigenvalues, then it is a consequence of 1.1.5 that each **Jordan block** J_i is 1×1; that is, the Jordan form is diagonal. But distinctness of the eigenvalues is by no means necessary in order to have a diagonal canonical form, as has already been seen in the case of a real symmetric matrix.

Although it is necessary for the matrix A to have multiple eigenvalues in order to have a nondiagonal canonical form, the algebraic multiplicity of an eigenvalue does not determine the structure of the Jordan form. For example, suppose that A is 4×4 and has the eigenvalue 2 with multiplicity 4. Then the possible Jordan forms of A, up to permutations of the blocks, are

$$\begin{bmatrix} 2 & & & \\ & 2 & & \\ & & 2 & \\ & & & 2 \end{bmatrix}, \begin{bmatrix} 2 & & & \\ & 2 & & \\ & & 2 & 1 \\ & & & 2 \end{bmatrix}, \begin{bmatrix} 2 & 1 & & \\ & 2 & & \\ & & 2 & 1 \\ & & & 2 \end{bmatrix}$$

$$\begin{bmatrix} 2 & 1 & & \\ & 2 & 1 & \\ & & 2 & \\ & & & 2 \end{bmatrix}, \begin{bmatrix} 2 & 1 & & \\ & 2 & 1 & \\ & & 2 & 1 \\ & & & 2 \end{bmatrix}.$$

It is easy to see (E1.1.10) that a matrix of the form

$$
J = \begin{bmatrix} \lambda & 1 & & & \\ & \cdot & \cdot & & \\ & & \cdot & 1 & \\ & & & \cdot & \\ & & & & \lambda \end{bmatrix} \tag{12}
$$

has precisely one linearly independent eigenvector. It follows from this that, in general, a matrix A has precisely as many linearly independent eigenvectors as it has Jordan blocks in its canonical form (E1.1.10). These eigenvectors are certain columns of the matrix P of the transformation $P^{-1}AP$ to canonical form. If A has fewer than n linearly independent eigenvectors, then the other columns of P are called **generalized eigenvectors** or **principal vectors**.

These principal vectors may be characterized algebraically in the following way. Consider again the matrix (12), and denote by $\mathbf{e}_1 = (1, 0, \ldots, 0)^T$, $\ldots, \mathbf{e}_n = (0, \ldots, 0, 1)^T$, the unit coordinate vectors. Clearly, $\mathbf{e}_1$ is an eigenvector of (12). Furthermore, it is a trivial calculation to see that

$$J\mathbf{e}_k = \lambda\mathbf{e}_k + \mathbf{e}_{k-1}, \qquad k = 2, \ldots, n.$$

Hence, since $(J - \lambda I)\mathbf{e}_1 = \mathbf{0}$,

$$(J - \lambda I)^{k-1}\mathbf{e}_k = \mathbf{e}_1, \qquad (J - \lambda I)^k\mathbf{e}_k = \mathbf{0}, \qquad k = 1, \ldots, n.$$

These relations characterize principal vectors for a general matrix. That is, $\mathbf{x}$ is a principal vector of **degree k** of $A \in L(C^n)$ if

$$(A - \lambda I)^k\mathbf{x} = \mathbf{0}, \qquad (A - \lambda I)^{k-1}\mathbf{x} \neq \mathbf{0}.$$

By this definition, an eigenvector is a principal vector of degree 1. We will also say that $\mathbf{x}$ is a principal vector of **maximal degree** if $\mathbf{x}$ is of degree m where m is the dimension of the largest Jordan block associated with the corresponding eigenvalue.

Note that, just as eigenvectors are determined only up to scalar multiples, principal vectors are determined only up to scalar multiples of themselves as well as linear combinations of principal vectors of lower degree.

We end this section with a note on real and complex matrices. In most of our considerations in the sequel we will be interested only in problems in which matrices are real; for example, a system of linear equations with a real coefficient matrix A. However, the analysis of various methods for the solution of such problems may require us to consider the eigenvalues and, more usually, Jordan canonical form of A. But even if A is real, its eigenvalues and eigenvectors may well be complex so that we are forced to shift attention from the real space R^n to complex space C^n. Hence, we will, on occasion, be bringing in C^n, sometimes tacitly, in the midst of our analysis.

EXERCISES

E1.1.1   Let $A \in L(R^n)$ be orthogonal. Show that all eigenvalues of A have absolute value unity.

E1.1.2   Let $A \in L(R^n)$ be symmetric. Show that all eigenvalues of A are real.

E1.1.3   A matrix $A \in L(R^n)$ is **skew-symmetric** if $A^T = -A$. Show that all eigenvalues of a skew-symmetric matrix are imaginary.

E1.1.4   Let $A \in L(R^n)$ be symmetric positive semidefinite. Show that all eigenvalues of A are nonnegative. Show also that all eigenvalues of A are positive if A is positive definite.

E1.1.5   Let $A \in L(R^n)$ be symmetric and let $\mathbf{x}$ and $\mathbf{y}$ be eigenvectors corresponding to distinct eigenvalues. Show that $\mathbf{x}$ and $\mathbf{y}$ are orthogonal.

E1.1.6   If λ and $\mathbf{x}$ are an eigenvalue and corresponding eigenvector of $A \in L(C^n)$, show that $A^k\mathbf{x} = \lambda^k\mathbf{x}$ for any positive integer k. Hence conclude that if A has eigenvalues $\lambda_1, \ldots, \lambda_n$ and eigenvectors $\mathbf{x}_1, \ldots, \mathbf{x}_n$, then A^k has eigenvalues $\lambda_1^k, \ldots, \lambda_n^k$ and eigenvectors $\mathbf{x}_1, \ldots, \mathbf{x}_n$. Show that the same result holds for negative integers k if A is nonsingular.

E1.1.7   With the notation of E1.1.6, show that $A - cI$ has eigenvalues $\lambda_1 - c, \ldots, \lambda_n - c$. More generally, if $p(\lambda) = a_n \lambda^n + \cdots + a_0$, define the matrix polynomial $p(A)$ by $p(A) = a_n A^n + \cdots + a_0 I$. Show that $p(A)$ has eigenvalues $p(\lambda_i)$, $i = 1, \ldots, n$ and eigenvectors $\mathbf{x}_1, \ldots, \mathbf{x}_n$.

E1.1.8   Compute the eigenvalues, eigenvectors, and principal vectors for

$$A = \begin{bmatrix} 1 & 2 \\ 4 & 3 \end{bmatrix}, \qquad A = \begin{bmatrix} 1 & 0 & 0 \\ -1 & 0 & 1 \\ -1 & -1 & 2 \end{bmatrix}.$$

E1.1.9   Use the Jordan canonical form theorem to show that $\det A = \lambda_1 \lambda_2 \cdots \lambda_n$, where $\lambda_1, \ldots, \lambda_n$ are the eigenvalues of A. Can you give a proof not using the Jordan form or 1.1.6?

E1.1.10 Show that the matrix of (12) has precisely one linearly independent eigenvector. Conclude from this that a matrix $A \in L(C^n)$ has precisely as many linearly independent eigenvectors as it has Jordan blocks in its canonical form.

E1.1.11 Compute the powers of the $n \times n$ matrix (12). Show, in particular, that

$$J^k = \begin{bmatrix} \lambda^k & k\lambda^{k-1} & \binom{k}{2}\lambda^{k-2} & \cdots & \binom{k}{n-1}\lambda^{k-n+1} \\ & \lambda^k & k\lambda^{k-1} & \cdots & \vdots \\ & & \cdot & \cdot & \cdot & \\ & & & \cdot & \cdot & \\ & & & & \cdot & k\lambda^{k-1} \\ & & & & & \lambda^k \end{bmatrix}$$

for $k \geq n$.

E1.1.12 Give a proof of Schur's theorem 1.1.6 along the lines of 1.1.3.

1.2 VECTOR NORMS

In this section, we will review the basic properties of norms on R^n and C^n.

1.2.1 Definition   A norm on R^n (or C^n) is a real-valued function $\|\cdot\|$ satisfying

(a) $\|\mathbf{x}\| \geq 0, \forall \mathbf{x}; \|\mathbf{x}\| = 0$ only if $\mathbf{x} = \mathbf{0}$;

(b) $\|\alpha\mathbf{x}\| = |\alpha| \, \|\mathbf{x}\|$ for all $\mathbf{x}$ and all scalars α;

(c) $\|\mathbf{x} + \mathbf{y}\| \leq \|\mathbf{x}\| + \|\mathbf{y}\|, \forall \mathbf{x}, \mathbf{y}$.

The main examples of norms with which we shall be concerned are

$$\|\mathbf{x}\|_2 = \left(\sum_{i=1}^{n} |x_i|^2 \right)^{1/2} \qquad \text{(the } l_2 \text{ or Euclidean norm)} \qquad (1)$$

$$\|\mathbf{x}\|_1 = \sum_{i=1}^{n} |x_i| \qquad \text{(the } l_1 \text{ or sum norm)} \qquad (2)$$

$$\|\mathbf{x}\|_\infty = \max_{1 \leq i \leq n} |x_i| \qquad \text{(the } l_\infty \text{ or max norm)}. \qquad (3)$$

The norms (1) and (2) are special cases of the general class of l_p norms

$$\|\mathbf{x}\|_p = \left(\sum_{i=1}^n |x_i|^p \right)^{1/p} \tag{4}$$

where $p \in [1, \infty)$. The norm (3) is the limiting case of (4) as $p \to \infty$. (See E1.2.2.)

Another important class of norms are the so-called **elliptic norms** defined by

$$\|\mathbf{x}\| = (\mathbf{x}^T B \mathbf{x})^{1/2} \tag{5}$$

where $B \in L(R^n)$ is an arbitrary symmetric positive definite matrix. The definitions (1)–(4) hold, of course, on either R^n or C^n; (5) holds on C^n if $\mathbf{x}^T$ is replaced by $\mathbf{x}^H$, and B is hermitian and positive definite.

It is easy to show that (2) and (3) satisfy the axioms of 1.2.1 (E1.2.1). The verification that (5) defines a norm—including the special case $B = I$ which is (1)—is not quite so obvious. We will prove that (5) is in fact an "inner product norm."

1.2.2 **Definition** A real-valued mapping $(\cdot, \cdot)$ on $R^n \times R^n$ is an **inner product** if

(a) $(\mathbf{x}, \mathbf{x}) \geq 0, \forall \mathbf{x}; (\mathbf{x}, \mathbf{x}) = 0$ only if $\mathbf{x} = \mathbf{0}$;
(b) $(\alpha \mathbf{x}, \mathbf{y}) = \alpha(\mathbf{x}, \mathbf{y})$ for all $\mathbf{x}, \mathbf{y} \in R^n$ and scalars α;
(c) $(\mathbf{x}, \mathbf{y}) = (\mathbf{y}, \mathbf{x}), \forall \mathbf{x}, \mathbf{y}$;
(d) $(\mathbf{x} + \mathbf{z}, \mathbf{y}) = (\mathbf{x}, \mathbf{y}) + (\mathbf{z}, \mathbf{y}), \forall \mathbf{x}, \mathbf{y}, \mathbf{z}$.

The same axioms define an inner product on $C^n \times C^n$ provided that the mapping is allowed to be complex-valued, and (c) is replaced by $(\mathbf{x}, \mathbf{y}) = \overline{(\mathbf{y}, \mathbf{x})}$.

For any inner product, a norm may be defined by

$$\|\mathbf{x}\| = (\mathbf{x}, \mathbf{x})^{1/2}. \tag{6}$$

The verification of axioms (a) and (b) of 1.2.1 is immediate. In order to prove the triangle inequality (c), we first prove the **Cauchy–Schwarz inequality**:

$$|(\mathbf{x}, \mathbf{y})|^2 \leq (\mathbf{x}, \mathbf{x})(\mathbf{y}, \mathbf{y}). \tag{7}$$

This results from the observation that the polynomial

$$p(\alpha) = \alpha^2(\mathbf{x}, \mathbf{x}) + 2\alpha(\mathbf{x}, \mathbf{y}) + (\mathbf{y}, \mathbf{y}) = (\alpha\mathbf{x} + \mathbf{y}, \alpha\mathbf{x} + \mathbf{y})$$

is always nonnegative and hence its discriminant, $(\mathbf{x}, \mathbf{y})^2 - (\mathbf{x}, \mathbf{x})(\mathbf{y}, \mathbf{y})$, is nonpositive. Using (7), the triangle inequality follows from

$$\|\mathbf{x} + \mathbf{y}\|^2 = (\mathbf{x}, \mathbf{x}) + 2(\mathbf{x}, \mathbf{y}) + (\mathbf{y}, \mathbf{y}) \leq (\mathbf{x}, \mathbf{x}) + 2(\mathbf{x}, \mathbf{x})^{1/2}(\mathbf{y}, \mathbf{y})^{1/2} + (\mathbf{y}, \mathbf{y})$$
$$= (\|\mathbf{x}\| + \|\mathbf{y}\|)^2.$$

Hence (6) is a norm. It is easily verified that if $B \in L(R^n)$ is symmetric positive definite, then $\mathbf{x}^T B \mathbf{y}$ is an inner product and therefore (5) is also a norm.

The verification that (4) is a norm for arbitrary $p \in [1, \infty)$ is left to E1.2.3.

For any norm, the set $\{\mathbf{x} : \|\mathbf{x}\| \leq 1\}$ is called the **unit ball** while the surface $\{\mathbf{x} : \|\mathbf{x}\| = 1\}$ is the **unit sphere**. These sets are depicted in Figure 1.2.1 for the norms (1)–(5).

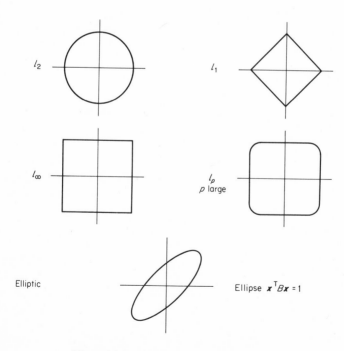

Figure 1.2.1 Unit balls of several norms.

Although the l_1, l_2, and l_∞ norms are those most common in practice, it is often useful, especially for theoretical purposes, to generate other norms by the following mechanism.

1.2.3 Let $\|\cdot\|$ be an arbitrary norm on R^n (or C^n) and P an arbitrary nonsingular $n \times n$ real (or complex) matrix. Then $\|\mathbf{x}\|' = \|P\mathbf{x}\|$ defines a norm on R^n (or C^n).

The proof is a simple verification of the axioms of 1.2.1 and is left to E1.2.4.

Given a sequence $\{\mathbf{x}_k\}$ of vectors, convergence may be defined in terms of the components of the vectors and this is obviously equivalent to convergence in the l_∞ norm; that is, $\mathbf{x}_k \to \mathbf{x}$ as $k \to \infty$ if and only if

$$\lim_{k \to \infty} \|\mathbf{x}_k - \mathbf{x}\|_\infty = 0.$$

However, it is an interesting fact that convergence in some norm implies convergence in any other norm. This is a consequence of the following important result.

1.2.4 **(Norm Equivalence Theorem)** Let $\|\cdot\|$ and $\|\cdot\|'$ be any two norms on R^n (or C^n). Then there are constants $c_2 \geq c_1 > 0$ such that

$$c_1 \|\mathbf{x}\| \leq \|\mathbf{x}\|' \leq c_2 \|\mathbf{x}\|, \qquad \forall \mathbf{x}. \tag{8}$$

Proof: It suffices to assume that $\|\cdot\|'$ is the l_2 norm. For if the two relations

$$d_1 \|\mathbf{x}\| \leq \|\mathbf{x}\|_2 \leq d_2 \|\mathbf{x}\|, \qquad d_1' \|\mathbf{x}\|' \leq \|\mathbf{x}\|_2 \leq d_2' \|\mathbf{x}\|'$$

both hold, then (8) holds with $c_1 = d_1/d_2'$ and $c_2 = d_2/d_1'$.

Let $\mathbf{e}_i$, $i = 1, \ldots, n$, be the coordinate vectors, Then for arbitrary $\mathbf{x}$

$$\mathbf{x} = \sum_{i=1}^{n} x_i \mathbf{e}_i$$

so that by the Cauchy–Schwarz inequality

$$\|\mathbf{x}\| \leq \sum_{i=1}^{n} |x_i| \, \|\mathbf{e}_i\| \leq \beta \|\mathbf{x}\|_2, \qquad \beta = \left(\sum_{i=1}^{n} \|\mathbf{e}_i\|^2 \right)^{1/2}. \tag{9}$$

Hence, the left-hand inequality of (8) holds with $c_1 = \beta^{-1}$. Also, using the result of E1.2.5,

$$|\,\|\mathbf{x}\| - \|\mathbf{y}\|\,| \le \|\mathbf{x} - \mathbf{y}\| \le \beta\,\|\mathbf{x} - \mathbf{y}\|_2$$

which shows that $\|\cdot\|$ is continuous with respect to the l_2 norm. Therefore, since the unit sphere $S = \{\mathbf{x} : \|\mathbf{x}\|_2 = 1\}$ is compact, $\|\cdot\|$ is bounded away from zero on S; that is, $\|\mathbf{x}\| \ge \alpha > 0$ for some $\alpha > 0$ and all $\mathbf{x} \in S$. Hence for arbitrary $\mathbf{x}$

$$\|\mathbf{x}\| = \|\mathbf{x}\|_2 \left\|\frac{\mathbf{x}}{\|\mathbf{x}\|_2}\right\| \ge \alpha\,\|\mathbf{x}\|_2$$

so that the right-hand inequality of (8) holds with $c_2 = \alpha^{-1}$. $\$\$\$$

EXERCISES

E1.2.1   Verify that (2) and (3) satisfy the axioms of 1.2.1.

E1.2.2   Let $\mathbf{x}$ be fixed but arbitrary. Show that (3) is the limit of (4) as $p \to \infty$.

E1.2.3   For any $p \in (1, \infty)$, verify the **Hölder inequality**

$$\left|\sum_{i=1}^{n} x_i y_i\right| \le \left(\sum_{i=1}^{n} |x_i|^p\right)^{1/p} \left(\sum_{i=1}^{n} |y_i|^q\right)^{1/q}$$

where $p^{-1} + q^{-1} = 1$. Use this inequality to show that (4) is a norm. (*Hint:* For any $a > 0$, $b > 0$, $\alpha > 0$, $\beta > 0$ with $\alpha + \beta = 1$, show that

$$a^\alpha b^\beta \le \alpha a + \beta b$$

and apply this with $\alpha = p^{-1}$, $\beta = q^{-1}$, $a = |x_i|^p / \sum |x_i|^p$, $b = |y_i|^q / \sum |y_i|^q$.)

E1.2.4   Prove Theorem 1.2.3.

E1.2.5   Let $\|\cdot\|$ be any norm on R^n (or C^n). Show that

$$|\,\|\mathbf{x}\| - \|\mathbf{y}\|\,| \le \|\mathbf{x} - \mathbf{y}\|, \qquad \forall \mathbf{x}, \mathbf{y}.$$

1.3 MATRIX NORMS

Corresponding to any vector norm on R^n there is a natural induced matrix (or operator) norm.

1.3.1 Definition Let $\|\cdot\|$, $\|\cdot\|'$ be arbitrary norms on R^n and R^m respectively. Then for any $A \in L(R^n, R^m)$, the quantity

$$\|A\| = \max_{\mathbf{x} \neq \mathbf{0}} \frac{\|A\mathbf{x}\|'}{\|\mathbf{x}\|} = \max_{\|\mathbf{x}\| = 1} \|A\mathbf{x}\|' \tag{1}$$

is the corresponding matrix norm. The analogous definition holds for C^n and C^m.

Note that (1) implies that $\|A\mathbf{x}\|' \leq \|A\| \|\mathbf{x}\|$ for any $\mathbf{x} \in R^n$.

The following result shows that (1) defines a norm on the nm-dimensional linear space $L(R^n, R^m)$.

1.3.2 (a) $\|A\| \geq 0$; $\|A\| = 0$ only if $A = 0$;

(b) $\|\alpha A\| = |\alpha| \|A\|$;

(c) $\|A + B\| \leq \|A\| + \|B\|$.

An important additional property of matrix norms is that

$$\|AB\| \leq \|A\| \|B\| \tag{2}$$

whenever $A \in L(R^n, R^m)$ and $B \in L(R^p, R^n)$ (provided, of course, that R^n has the same norm as the domain of A and the range of B). The proofs of (2) and 1.3.2 are immediate consequences of 1.3.1 and the properties of vector norms and are left to E1.3.1.

As with vectors, the convergence of a sequence of matrices may be defined componentwise or, equivalently, in terms of any matrix norm. That is, we write $A_k \to A$ as $k \to \infty$ if in some norm $\|A_k - A\| \to 0$ as $k \to \infty$. By 1.2.4, convergence in some norm implies convergence in any norm.

The geometric interpretation of a matrix norm is that $\|A\|$ is the maximum length of a unit vector after transformation by A; this is depicted in Figure 1.3.1 for the l_2 norm.

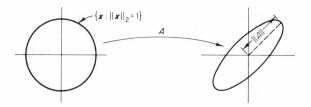

Figure 1.3.1

As Figure 1.3.1 indicates for two dimensions, $\|A\|$ in the Euclidean norm is the length of the major axis of the ellipse $\{A\mathbf{x}: \|\mathbf{x}\|_2 = 1\}$. This is true in any number of dimensions and the length of the major axis is given by (3). We first prove the following generally useful lemma.

1.3.3   Let $B \in L(R^n)$ be symmetric with eigenvalues $\lambda_1 \leq \cdots \leq \lambda_n$. Then

$$\lambda_1 \mathbf{x}^T\mathbf{x} \leq \mathbf{x}^T B\mathbf{x} \leq \lambda_n \mathbf{x}^T\mathbf{x}, \qquad \forall \mathbf{x} \in R^n.$$

Proof: By 1.1.3, there is an orthogonal matrix P such that

$$P^T BP = \text{diag}(\lambda_1, \ldots, \lambda_n).$$

Hence, with $\mathbf{y} = P^T\mathbf{x}$,

$$\mathbf{x}^T B\mathbf{x} = \mathbf{y}^T P^T BP\mathbf{y} = \sum_{i=1}^{n} \lambda_i y_i^2 \leq \lambda_n \mathbf{y}^T\mathbf{y} = \lambda_n \mathbf{x}^T\mathbf{x}.$$

The other inequality is proved analogously. $$$

The previous result also holds for hermitian matrices if $\mathbf{x}^T$ is replaced by $\mathbf{x}^H$. Similarly, the following theorem holds for complex matrices if A^T is replaced by A^H.

1.3.4   Let $A \in L(R^n)$. Then

$$\|A\|_2 \equiv \max_{\|x\|_2=1} \|A\mathbf{x}\|_2 = [\rho(A^T A)]^{1/2}. \tag{3}$$

Proof: Set $\mu = [\rho(A^T A)]^{1/2}$. Then for any $x \in R^n$, 1.3.3 shows that

$$\|A\mathbf{x}\|_2^2 = \mathbf{x}^T A^T A\mathbf{x} \leq \mu^2 \mathbf{x}^T\mathbf{x}$$

so that $\|A\|_2 \leq \mu$. On the other hand, if $\mathbf{u}$ is an eigenvector of $A^{\mathrm{T}}A$ corresponding to μ^2, then

$$\mathbf{u}^{\mathrm{T}}A^{\mathrm{T}}A\mathbf{u} = \mu^2\mathbf{u}^{\mathrm{T}}\mathbf{u}$$

which shows that equality holds in (3). $\$\$\$$

We note that the eigenvalues of $A^{\mathrm{T}}A$ are called the **singular values** of A. Since $A^{\mathrm{T}}A$ is positive semidefinite (EI.3.4) all of A's singular values are nonnegative. The number of positive singular values is precisely the rank of A (EI.3.5).

It is usually difficult to compute $\|A\|$ explicitly for an arbitrary norm, and even for the l_2 norm as 1.3.4 shows. For the l_1 and l_∞ norms, however, it is immediate.

1.3.5   Let $A \in L(R^n)$. Then

$$\|A\|_\infty \equiv \max_{\|\mathbf{x}\|_\infty = 1} \|A\mathbf{x}\|_\infty = \max_{1 \leq i \leq n} \sum_{j=1}^{n} |a_{ij}| \tag{4}$$

and

$$\|A\|_1 \equiv \max_{\|\mathbf{x}\|_1 = 1} \|A\mathbf{x}\|_1 = \max_{1 \leq j \leq n} \sum_{i=1}^{n} |a_{ij}|. \tag{5}$$

Proof: Consider first the l_1 norm; then, for any $\mathbf{x} \in R^n$,

$$\|A\mathbf{x}\|_1 = \sum_{i=1}^{n} \left| \sum_{j=1}^{n} a_{ij} x_j \right| \leq \sum_{i=1}^{n} \sum_{j=1}^{n} |a_{ij}| \, |x_j| = \sum_{j=1}^{n} |x_j| \sum_{i=1}^{n} |a_{ij}|$$

$$\leq \max_{1 \leq j \leq n} \sum_{i=1}^{n} |a_{ij}| \, \|\mathbf{x}\|_1.$$

To show that there is some $\mathbf{x} \neq \mathbf{0}$ for which equality is attained in (5), let k be such that

$$\max_{1 \leq j \leq n} \sum_{i=1}^{n} |a_{ij}| = \sum_{i=1}^{n} |a_{ik}|.$$

Then, if $\mathbf{e}_k$ is the kth coordinate vector and $\mathbf{a}_k$ the kth column of A, we have

$$\|A\mathbf{e}_k\|_1 = \|\mathbf{a}_k\|_1 = \sum_{i=1}^{n} |a_{ik}|.$$

The proof for $\|A\|_\infty$ is similar. In this case, the maximum is taken on for the vector defined by

$$x_i = \begin{cases} a_{ki}/|a_{ki}|, & a_{ki} \neq 0 \\ 1, & a_{ki} = 0 \end{cases}$$

where, again, k is chosen so that the maximum in (4) is achieved. $\$\$\$$

We note that (4) and (5) obviously also hold for complex matrices. The quantity of (4) is often called the "maximum row sum" of A while (5) is the "maximum column sum."

An important special case of 1.3.4 is that if $A \in L(R^n)$ is symmetric, then

$$\|A\|_2 = [\rho(A^2)]^{1/2} = [\rho(A)^2]^{1/2} = \rho(A)$$

where we have used the result of E1.1.6.

The question thus arises as to the general relation of $\rho(A)$ and $\|A\|$. If λ is any eigenvalue of A and $\mathbf{x} \neq \mathbf{0}$ a corresponding eigenvector, then in any norm (perhaps necessarily on C^n), we have

$$\|A\| \geq \|A\mathbf{x}\|/\|\mathbf{x}\| = |\lambda|.$$

That is,

$$\rho(A) \leq \|A\|.$$

On the other hand, for the matrix

$$A = \begin{pmatrix} 0 & \alpha \\ 0 & 0 \end{pmatrix} \tag{6}$$

we have $\rho(A) = 0$ and $\|A\|_1 = |\alpha|$ so that, in general, the difference between $\|A\|$ and $\rho(A)$ may be arbitrarily large. In some sense, this is because of the choice of the wrong norm as the following important result shows.

1.3.6 Let $A \in L(C^n)$. Then given any $\varepsilon > 0$ there is a norm on C^n such that

$$\|A\| \leq \rho(A) + \varepsilon.$$

Proof: Let $A = PJP^{-1}$, where J is the Jordan form of A, and let

$$D = \operatorname{diag}(1, \varepsilon, \varepsilon^2, \ldots, \varepsilon^{n-1}).$$

Then it is easy to see that $\hat{J} = D^{-1}JD$ is exactly the same as J except that every off-diagonal 1 in J is replaced by ε. Hence

$$\|\hat{J}\|_\infty \le \rho(A) + \varepsilon.$$

Set $Q = PD$ and define $\|\mathbf{x}\| = \|Q^{-1}\mathbf{x}\|_\infty$; this is a norm by 1.2.3. Then

$$\|A\| = \max_{\|\mathbf{x}\|=1} \|A\mathbf{x}\| = \max_{\|Q^{-1}\mathbf{x}\|_\infty=1} \|Q^{-1}A\mathbf{x}\|_\infty = \max_{\|\mathbf{y}\|_\infty=1} \|Q^{-1}AQ\mathbf{y}\|_\infty$$

$$= \max_{\|\mathbf{y}\|_\infty=1} \|\hat{J}\mathbf{y}\|_\infty = \|\hat{J}\|_\infty \le \rho(A) + \varepsilon. \quad \$\$\$$$

Theorem 1.3.6 gives the best possible result in the sense that it is not always possible to find a norm for which

$$\rho(A) = \|A\|. \tag{7}$$

This is shown by the example (6) since for any norm we must have $\|A\| > 0$ provided that $\alpha \ne 0$. In fact, for (7) to hold it is necessary and sufficient that A belong to the following class of matrices.

1.3.7   Definition   A matrix $A \in L(C^n)$ is of **class M** if for every eigenvalue λ such that $|\lambda| = \rho(A)$ every Jordan block associated with λ is 1×1.

Equivalently, we can say that A is of class M if and only if A is similar to a matrix of the form

$$\begin{bmatrix} A_1 & 0 \\ 0 & A_2 \end{bmatrix}$$

where A_1 is diagonal, $\rho(A_1) = \rho(A)$, and $\rho(A_2) < \rho(A)$.

1.3.8   Let $A \in L(C^n)$. Then there is a norm on C^n such that (7) holds if and only if A is of class M.

Proof: For the sufficiency, we modify slightly the proof of 1.3.6 by assuming that $\varepsilon > 0$ is chosen so small that $|\lambda| + \varepsilon < \rho(A)$, where λ is any eigenvalue of A such that $|\lambda| < \rho(A)$. Then it is easy to see that $\|\hat{J}\|_\infty = \rho(A)$. For the converse, assume that $\|A\| = \rho(A)$ for some norm, but that there is an $m \times m$ Jordan block, $m \ge 2$, associated with an eigenvalue λ such that $|\lambda| = \rho(A)$. Clearly, we may assume that $\lambda \ne 0$ for otherwise we would have $\|A\| = 0$ which implies that $A = 0$. Hence it suffices to consider a Jordan block

$$J = \begin{bmatrix} \lambda & 1 & & & \\ & \ddots & \ddots & & \\ & & \ddots & 1 \\ & & & \ddots & \\ & & & & \lambda \end{bmatrix}, \qquad \lambda \neq 0$$

and to show that $\|J\| = |\lambda|$ is not possible in any norm. If we assume that $\|J\| = |\lambda|$ and set $\hat{J} = \lambda^{-1}J$ then, clearly, $\|\hat{J}\| = 1$. But a direct computation† shows that $\hat{J}^k \mathbf{e}_2 = (k/\lambda, 1, 0, \ldots, 0)^{\mathrm{T}}$ so that $\|\hat{J}^k \mathbf{e}_2\| \to \infty$ as $k \to \infty$. This contradicts $\|\hat{J}\| = 1$. $$$

As an important corollary of 1.3.6 and 1.3.8 we obtain the following result on powers of a matrix.

1.3.9 Let $A \in L(C^n)$. Then $\lim_{k \to \infty} A^k = 0$ if and only if $\rho(A) < 1$. Moreover, $\|A^k\|$ is bounded as $k \to \infty$ if and only if $\rho(A) < 1$, or $\rho(A) = 1$ and A is of class M.

Proof: If $\rho(A) < 1$, then by 1.3.6 we may choose a norm on C^n such that $\|A\| < 1$. Hence

$$\|A^k\| \leq \|A\|^k \to 0 \qquad \text{as} \quad k \to \infty.$$

Conversely, suppose that $\rho(A) \geq 1$ and let λ be some eigenvalue such that $|\lambda| \geq 1$. If $\mathbf{x}$ is a corresponding eigenvector, then

$$\|A^k \mathbf{x}\| = \|\lambda^k \mathbf{x}\| \geq \|\mathbf{x}\|$$

which implies that $\|A^k\| \geq 1$ for all k.

For the second part, we have already shown that $A^k \to 0$ as $k \to \infty$ if $\rho(A) < 1$. If $\rho(A) = 1$ and A is of class M, then 1.3.8 ensures that we may find a norm such that $\|A\| = 1$. Hence $\|A^k\| \leq 1$ for all k. Conversely, suppose that $\{A^k\}$ is bounded. Then clearly $\rho(A) \leq 1$. Suppose that $\rho(A) = 1$. Then an argument precisely as in 1.3.8 shows that any Jordan block J such that $\rho(J) = 1$ is 1×1. Hence, A is of class M. $$$

† Alternatively, use J^k as given by E.1.1.11.

We end this section with another important corollary of 1.3.6, which is the matrix analogue of the geometric series

$$\frac{1}{1-\alpha} = 1 + \alpha + \alpha^2 + \cdots, \qquad |\alpha| < 1.$$

1.3.10   (Neumann Lemma)   Let $B \in L(C^n)$ with $\rho(B) < 1$. Then $(I - B)^{-1}$ exists and

$$(I - B)^{-1} = \lim_{k \to \infty} \sum_{i=0}^{k} B^i.$$

Proof: The eigenvalues of $I - B$ are $1 - \lambda_i$, $i = 1, \ldots, n$, where $\lambda_1, \ldots, \lambda_n$ are the eigenvalues of B. Since $\rho(B) < 1$, it follows that $-IB$ has no eigenvalue equal to zero and, hence, is nonsingular. Now consider the identity

$$(I - B)(I + B + \cdots + B^k) = I - B^{k+1}$$

or

$$I + B + \cdots + B^k = (I - B)^{-1} - (I - B)^{-1}B^{k+1}.$$

By 1.3.9 the second term on the right tends to zero as $k \to \infty$. $\$\$\$

EXERCISES

E1.3.1   Prove Theorem 1.3.2. and verify (2).

E1.3.2   Show that $\|I\| = 1$ in any norm.

E1.3.3   Let $A \in L(R^n)$ and define

$$\|A\| = \left(\sum_{i,j=1}^{n} a_{ij}^2 \right)^{1/2}.$$

Show that (a), (b), and (c) of 1.3.2 hold, but that this is not a norm in the sense of 1.3.1.

E1.3.4   Let $A \in L(R^n)$. Show that $A^T A$ is always symmetric positive semidefinite and positive definite if and only if A is nonsingular. State and prove the analogous result for complex matrices.

EI.3.5 Let $A \in L(R^n)$. Show that the number of positive eigenvalues of $A^T A$ is equal to the rank of A.

EI.3.6 Formulate and prove 1.3.3 and 1.3.4 for complex matrices. Formulate and prove 1.3.4 and 1.3.5 for nonsquare matrices.

EI.3.7 Compute $\|A\|_1$, $\|A\|_2$, and $\|A\|_\infty$ for the matrices of EI.1.8.

EI.3.8 Let $\|\cdot\|$ be an arbitrary norm on R^n (or C^n) and P an arbitrary non-singular real (or complex) matrix. Define the norm $\|x\|' = \|Px\|$. Show that $\|A\|' = \|PAP^{-1}\|$.

READING

The majority of the material in Section 1.1 may be found in any of the numerous books on linear algebra. In particular, the excellent treatise of Gantmacher [1953] presents the Jordan canonical form in great detail from both the algebraic (elementary divisor) and geometric (invariant subspace) points of view. See also the books by Faddeev and Faddeeva [1960] and Wilkinson [1965] for a review of many aspects of linear algebra in a form most suitable for numerical analysis, and, in particular, for a thorough study of norms. Householder [1964] develops the theory of norms in a more geometric way in terms of convex bodies.

Normed linear spaces—in particular, Banach and Hilbert spaces—play a central role in functional analysis, the study of infinite-dimensional linear spaces. For an introduction to this subject see, for example, Dieudonné [1969].

PART I

MATHEMATICAL
STABILITY AND ILL CONDITIONING

We shall begin our study of numerical analysis by examining certain typical classes of problems—linear equations, eigenvalues, differential equations, etc.—from the point of view of the following question: If small changes are made in the data of the problem, do large changes in the solution occur? If so, the solution or problem will be said to be **unstable** or **ill conditioned**.

The answer to the above question is of fundamental importance to numerical analysis in at least two distinct ways. First, if the problem is unstable, there may be no point in even attempting to obtain an approximate solution. For if the data of the problem are measured quantities, or otherwise known to only a limited precision, then even an exact solution to the problem (exact with respect to numerical analysis error) is quite likely to be meaningless. It may be much more important simply to detect the instability so that perhaps the problem can be reformulated into a more stable form. Second, there is usually no precise demarcation between stable and unstable problems but, rather, a continuum of possibilities. Some appropriate measure of this instability will usually enter into the error estimates for all three of the fundamental errors—discretization, convergence, and rounding. Therefore, it is desirable to have as precise a knowledge as possible of this "condition number."

CHAPTER 2

SYSTEMS OF LINEAR ALGEBRAIC EQUATIONS

2.1 BASIC ERROR ESTIMATES AND CONDITION NUMBERS

Consider the system of linear equations

$$AX = B \tag{1}$$

where $A \in L(R^n)$ and $X, B \in L(R^m, R^n)$; that is, A is a real $n \times n$ matrix, and X and B are real $n \times m$ matrices. (For simplicity, we consider only real matrices although most of our results hold equally well for complex matrices. Also, a complex system may always be converted to a real system of twice the dimension; see E2.1.1.) We will always assume that A is nonsingular so that (1) has a unique solution $X = A^{-1}B$. Important special cases of (1) are $m = n$ and $B = I$ so that $X = A^{-1}$, and $m = 1$; in the latter case, X and B may be considered as vectors $\mathbf{x}$ and $\mathbf{b}$ in R^n so that (1) becomes $A\mathbf{x} = \mathbf{b}$.

Our concern in this section will be to give estimates for the difference $X - Y$, where Y is the solution of the system

$$(A + E)Y = B + F. \tag{2}$$

The "perturbation" matrices E and F may arise in a variety of ways. It may be that the elements of A and B are measured quantities subject to some observation error. Hence the actual matrices with which we deal are $A + E$ and $B + F$ rather than the "true" A and B. Similarly, the elements of A and B may be the result of prior computations and, hence, contaminated by rounding error. The simplest example of this kind—but one of importance—occurs by rounding (or chopping) the elements of A and B

to a finite number of figures to enter them into a computer. Consider, for example, the **Hilbert matrix** of order n,

$$H_n = \begin{bmatrix} 1 & 1/2 & \cdots & 1/n \\ 1/2 & & & \\ \vdots & & & \vdots \\ 1/n & & \cdots & 1/(2n-1) \end{bmatrix} ; \tag{3}$$

if we were working with, say, a 27-bit binary machine, several of the elements of H_n—for example, $\frac{1}{3}$—would be truncated to 27 bits upon entering H_n in the machine.

Finally, we will see in Part IV that as a result of the analysis of the rounding error made in carrying out gaussian elimination on (1) that the computed solution satisfies exactly an equation of the form (2).

Now assume that $A + E$ is nonsingular and that Y satisfies (2). Then

$$X - Y = A^{-1}B - (A + E)^{-1}(B + F) = (A + E)^{-1}(EX - F), \tag{4}$$

which gives an exact, although somewhat clumsy, representation of the error $X - Y$. It will be convenient to recast this into a relative estimate using norms. From (1) we have

$$\|X\| \geq \|A\|^{-1}\|B\| \tag{5}$$

so that from (4) it follows that

$$\frac{\|X - Y\|}{\|X\|} \leq \|(A + E)^{-1}\|\left(\|E\| + \|A\|\frac{\|F\|}{\|B\|}\right). \tag{6}$$

It is now necessary to obtain an estimate for $\|(A + E)^{-1}\|$. The following basic result also gives a sufficient condition that $A + E$ be nonsingular.

2.1.1   (Perturbation Lemma, Banach Lemma)   Let $A \in L(R^n)$ be nonsingular. If $E \in L(R^n)$ and $\|A^{-1}\|\,\|E\| < 1$, then $A + E$ is nonsingular and

$$\|(A + E)^{-1}\| \leq \frac{\|A^{-1}\|}{1 - \|A^{-1}\|\,\|E\|}.$$

Proof: Set $B = -A^{-1}E$; then $\|B\| < 1$. It follows from the Neumann lemma 1.3.10 that $I - B$ is nonsingular and that

$$(I - B)^{-1} = \sum_{i=0}^{\infty} B^k.$$

Hence

$$\|(I - B)^{-1}\| \le \sum_{i=0}^{\infty} \|B\|^k = \frac{1}{1 - \|B\|} \le \frac{1}{1 - \|A\| \, \|E\|}.$$

But

$$A + E = A(I + A^{-1}E) = A(I - B)$$

so that $A + E$, as the product of nonsingular matrices, is nonsingular and

$$\|(A + E)^{-1}\| = \|(I - B)^{-1}A^{-1}\| \le \frac{\|A^{-1}\|}{1 - \|A^{-1}\| \, \|E\|}. \qquad \$\$\$$$

As a consequence of 2.1.1 together with (6) we may summarize our error estimate as follows.

2.1.2 (Error Estimate Theorem) Assume that the conditions of 2.1.1 hold for A and $E \in L(R^n)$ and that X and Y satisfy (1) and (2), respectively. Then

$$\frac{\|X - Y\|}{\|X\|} \le \frac{\|A^{-1}\| \, \|A\|}{1 - \|A^{-1}\| \, \|E\|} \left\{ \frac{\|E\|}{\|A\|} + \frac{\|F\|}{\|B\|} \right\}. \qquad (7)$$

Note that if $E = 0$, (7) reduces to simply

$$\frac{\|X - Y\|}{\|X\|} \le \|A^{-1}\| \, \|A\| \frac{\|F\|}{\|B\|} \qquad (E = 0);$$

that is, the relative error in X is bounded by the relative error in B times the factor $\|A^{-1}\| \, \|A\|$. This same factor plays a dominant role in (7) also, and becomes a useful measure of the stability of the solution X.

2.1.3 Definition Let $A \in L(R^n)$. Then

$$K(A) = \begin{cases} \|A\| \, \|A^{-1}\| & \text{if } A \text{ is nonsingular} \\ +\infty & \text{if } A \text{ is singular} \end{cases}$$

is the **condition number** of A (with respect to inversion and with respect to the particular norm used).

In terms of the condition number, (7) may be written as

$$\frac{\|X - Y\|}{\|X\|} \leq \frac{K(A)}{1 - K(A)(\|E\|/\|A\|)} \left\{ \frac{\|E\|}{\|A\|} + \frac{\|F\|}{\|B\|} \right\}. \tag{8}$$

Since, by assumption, $\alpha \equiv K(A)(\|E\|/\|A\|) = \|A^{-1}\| \, \|E\| < 1$, we may expand $(1 - \alpha)^{-1}$ by its geometric series

$$\frac{1}{1 - \alpha} = 1 + \alpha + \alpha^2 + \cdots$$

and conclude that, to a first-order approximation, the relative error in X is $K(A)$ times the relative errors in A and B.

As indicated in 2.1.3, $K(A)$ depends upon the norm; for the l_1, l_2, and l_∞ norms we will denote the corresponding condition number by $K_i(A)$, $i = 1, 2, \infty$.

In any norm, $1 \leq K(A) \leq \infty$ and for "large" values of $K(A)$, the matrix A is said to be **ill conditioned**. Here "large" must be interpreted in somewhat subjective terms. For example, if $K(A) = 100$, the estimate (8) shows that relative changes of 10^{-8} in the right-hand side may cause relative errors in the solution of 10^{-6}, which may or may not be considered to be a large degradation in the accuracy.

One limiting case of the condition number is when A is singular, while the other limiting case is when $K(A) = 1$ and then A is said to be **perfectly conditioned**. Note that whereas $K(A) = +\infty$ is norm independent, the class of perfectly conditioned matrices depends on the norm. For example, in the l_2 norm the perfectly conditioned matrices are precisely the class of scalar multiples of orthogonal matrices, but this is not necessarily true in other norms (see E2.1.5).

We return to the Hilbert matrix (3) which is perhaps the classical example of an ill-conditioned matrix. Suppose that $n = 6$ and let $\bar{H}_6$ denote the matrix with elements truncated to 27 binary digits of accuracy ($\bar{H}_6$ would be the matrix actually sitting in the memory of a 27-bit machine). Hence

$$\frac{\|E\|_\infty}{\|A\|_\infty} \equiv \frac{\|H_6 - \bar{H}_6\|_\infty}{\|H_6\|_\infty} \doteq 10^{-8}.$$

The exact solutions of the systems

$$H_6 \mathbf{x} = \mathbf{e}_1, \qquad \bar{H}_6 \bar{\mathbf{x}} = \mathbf{e}_1$$

are given by

$$\mathbf{x}^T = (36, -630, 3360, -7560, +7560, -2772)$$
$$\bar{\mathbf{x}}^T = (36.04 \cdots, -631.8 \cdots, 3374.8 \cdots, -7602.6 \cdots, 7610.3 \cdots,$$
$$-2792.8 \cdots)$$

so that

$$\frac{\|\mathbf{x} - \bar{\mathbf{x}}\|_\infty}{\|\mathbf{x}\|_\infty} \doteq 5 \cdot 10^{-4};$$

that is, the relative error in the solution is about $5 \cdot 10^4$ times the relative error in the coefficient matrix. We would guess from this that $K(H_6)$ is at least $5 \cdot 10^4$. Actually, it is even larger.

We note first that for any nonsingular symmetric matrix $A \in L(R^n)$, Theorem 1.3.4 together with E1.1.6 imply that

$$K_2(A) = |\lambda|_{max}/|\lambda|_{min} \tag{9}$$

where $|\lambda|_{max}$ and $|\lambda|_{min}$ are, respectively, the maximum and minimum absolute values of the eigenvalues of A. The eigenvalues of H_n for the first few values of n have been computed,† and the resulting condition numbers are tabulated in Table 1.

TABLE 1   Condition Numbers of the Hilbert Matrix

n	3	5	6	8
$K_2(H_n)$	$5 \cdot 10^2$	$5 \cdot 10^5$	$15 \cdot 10^6$	$15 \cdot 10^9$

For an arbitrary nonsymmetric matrix, (9) does not necessarily hold (see E2.1.10), but we always have

$$K(A) \geq |\lambda|_{max}/|\lambda|_{min} \tag{10}$$

in any norm. This follows from

$$\|A\| \geq \rho(A) = |\lambda|_{max}, \qquad \|A^{-1}\| \geq \rho(A^{-1}) = 1/|\lambda|_{min}.$$

Hence, if A has any small eigenvalues, or more precisely if the ratio of largest to smallest eigenvalue (in absolute value) is large, then A is ill conditioned although the converse does not necessarily hold (see E2.1.10 and E2.1.11).

† See, for example, R. Gregory and D. Karney, "A Collection of Matrices for Testing Computational Algorithms," Wiley (Interscience), New York, 1969.

We end this section by noting that Theorem 2.1.2 shows that the solution of the system $AX = B$ is a continuous function of the elements of A and B at any point for which A is nonsingular. It is important to realize that there are many computational problems for which this basic property of continuity does not hold; in some sense, discontinuity is the limiting case of ill conditioning. Consider, for example, the problem of computing the rank of A. This is, in general, an "impossible" problem because rank A is not necessarily a continuous function of A. For example, the matrix

$$A(\varepsilon) = \begin{bmatrix} 1 & 1 \\ 0 & \varepsilon \end{bmatrix}$$

has rank two for $\varepsilon \neq 0$ but rank $A(0) = 1$. This shows, in particular, that the problem of determining whether a matrix is singular is, in general, also "impossible."

EXERCISES

E2.1.1 Show that the complex linear system $AX = B$, where $A = A_1 + iA_2$, $B = B_1 + iB_2$, $X = X_1 + iX_2$, $i = (-1)^{1/2}$, may be solved by solving the real system

$$\begin{bmatrix} A_1 & -A_2 \\ A_2 & A_1 \end{bmatrix} \begin{bmatrix} X_1 \\ X_2 \end{bmatrix} = \begin{bmatrix} B_1 \\ B_2 \end{bmatrix}.$$

E2.1.2 Consider the system

$$A\mathbf{x} \equiv \begin{bmatrix} 1 & 1 \\ 1 & 1.01 \end{bmatrix} \begin{bmatrix} x_1 \\ x_2 \end{bmatrix} = \begin{bmatrix} 2 \\ 2.01 \end{bmatrix}$$

with exact solution $x_1 = x_2 = 1$, and the system

$$(A + E)\mathbf{y} \equiv \begin{bmatrix} 1 & 1 \\ 1 & 1.011 \end{bmatrix} \begin{bmatrix} y_1 \\ y_2 \end{bmatrix} = \begin{bmatrix} 2 \\ 2.01 \end{bmatrix}.$$

Compute $K_\infty(A)$. Compute $\|\mathbf{x} - \mathbf{y}\|_\infty / \|\mathbf{x}\|_\infty$ and the estimate for this from the error estimate theorem 2.1.2.

E2.1.3 Prove 2.1.1 and 2.1.2 for complex matrices.

E2.1.4 Calculate $K_1(A)$, $K_2(A)$, and $K_\infty(A)$ for

$$A = \begin{bmatrix} 1 & 2 \\ 3 & 4 \end{bmatrix}.$$

E2.1.5 Let $A \in L(R^n)$ be nonsingular. Show that $K_2(A) = 1$ if and only if

A is a (nonzero) scalar multiple of an orthogonal matrix. Show, however, that $K_\infty(A) \neq 1$ for the orthogonal matrix

$$\frac{1}{\sqrt{5}} \begin{bmatrix} 2 & -1 \\ 1 & 2 \end{bmatrix}.$$

E2.1.6 Let $A \in L(R^n)$ be nonsingular. Show that $K(A) = K(\alpha A)$ for any nonzero scalar α. Show that $K_2(A) = K_2(AU) = K_2(UA)$ for any orthogonal $U \in L(R^n)$.

E2.1.7 Let $D = \text{diag}(10^{-1}, \ldots, 10^{-1}) \in L(R^n)$. Compute det D and $K_2(D)$ as a function of n. Comment on the size of the condition number compared with the determinant.

E2.1.8 Let $A \in L(R^n)$ be nonsingular and $E = \alpha A$ for $|\alpha| < 1$. Show that the solutions of $Ax = b$ and $(A + E)y = b$ satisfy

$$\|x - y\| \leq |\alpha| \, \|x\|/(1 - |\alpha|).$$

[Note the absence of $K(A)$.]

E2.1.9 Consider the matrix $A = I + \alpha uu^T$ for some vector u with $u^T u = 1$. Show that the eigenvalues of A are 1, with multiplicity $n - 1$, and $1 + \alpha$. Hence, for an arbitrary number $N \geq 1$ choose α such that $K_2(A) = N$. (This gives a simple means of generating full matrices of arbitrary condition number.)

E2.1.10 Consider the matrices

$$A = \begin{bmatrix} 1 & -1 \\ 1 & -1.00001 \end{bmatrix}, \qquad B = \begin{bmatrix} 1 & -1 \\ -1 & 1.00001 \end{bmatrix}.$$

Show that the ratio of maximum to minimum eigenvalue is about 1 for A and about $4 \cdot 10^5$ for B. Show, however, that $K_2(A) = K_2(B)$. Conclude that the ratio of maximum to minimum eigenvalue is not a good condition number for nonsymmetric matrices. Is A well conditioned or ill conditioned?

E2.1.11 Consider the 100×100 matrix

$$A = \begin{bmatrix} 0.501 & & -1 & & & \\ & 0.502 & & -1 & & \\ & & \cdot & & \ddots & \\ & & & \cdot & & -1 \\ & & & & \cdot & 0.600 \end{bmatrix}.$$

Show that the first component of the solution of $A\mathbf{x} = \mathbf{e}_1$ is

$$x_1 = 1/(0.600 \times 0.599 \times \cdots \times 0.501) > (0.6)^{-100} > 10^{22}$$

and hence $K_\infty(A) > 10^{21}$ although $|\lambda|_{max}/|\lambda|_{min} \doteq 1$. (See Wilkinson [1965] for further discussion of this example.)

2.2 A POSTERIORI BOUNDS AND EIGENVECTOR COMPUTATIONS

If $A \in L(R^n)$ is ill conditioned, then small changes in the elements of A or B may produce large changes in the solutions of

$$AX = B. \tag{1}$$

Another consequence of ill conditioning is that it may be difficult to determine if an approximate solution Y of (1) is sufficiently accurate. The standard test of such an approximate solution is to compute the **residual matrix**

$$R = AY - B \tag{2}$$

and if the elements of R are "small" then one concludes that Y is satisfactory. In fact, the "smallness" of R is no guarantee that Y is close to the true solution X. (See E2.2.1.) As in the previous section, it is the condition number, $K(A) = \|A\| \, \|A^{-1}\|$, which again plays the crucial role.

2.2.1 (A Posteriori Error Estimate) Let $A \in L(R^n)$ be nonsingular and let X, Y, B, and $R \in L(R^n, R^m)$ satisfy (1) and (2). Then

$$Y - X = A^{-1}R, \tag{3}$$

and

$$\frac{\|Y - X\|}{\|X\|} \leq K(A) \frac{\|R\|}{\|B\|}. \tag{4}$$

Proof: Clearly

$$Y - X = A^{-1}(R + B) - A^{-1}B = A^{-1}R.$$

Then, since $\|X\| \geq \|A\|^{-1}\|B\|$,

$$\frac{\|Y - X\|}{\|X\|} \leq \frac{\|A^{-1}\| \, \|R\|}{\|A\|^{-1}\|B\|}$$

which is (4). $\$\$\$

Note that in the special case $B = I$, (4) reduces to

$$\frac{\| Y - A^{-1} \|}{\| A^{-1} \|} \le K(A) \| R \|.$$

It is important to realize that there are problems in which the coefficient matrix of the linear system is necessarily ill conditioned, but that this does not lead to erroneous results. Consider the computation of an eigenvector of A once an approximate eigenvalue $\bar{\lambda}$ has been obtained. Although, by definition, the eigenvector is a nonzero solution of the homogeneous system $(A - \lambda I)\mathbf{x} = 0$, it is much better in practice to attempt to solve the inhomogeneous system

$$(A - \bar{\lambda} I)\mathbf{x} = \mathbf{b} \tag{5}$$

for some "almost arbitrary" right-hand side $\mathbf{b}$.

In order to understand why this procedure is valid, and also what restrictions need to be placed on $\mathbf{b}$, we will analyze the solution of (5) in terms of the eigensystem of A. Assume, for simplicity, that A has n linearly independent eigenvectors $\mathbf{u}_1, \ldots, \mathbf{u}_n$, corresponding to the eigenvalues $\lambda_1, \ldots, \lambda_n$. Then $\mathbf{b}$ may be expanded in terms of the eigenvectors as

$$\mathbf{b} = \sum_{i=1}^{n} \alpha_i \mathbf{u}_i.$$

Now assume that $\bar{\lambda} \doteq \lambda_1$ but that $\bar{\lambda} \ne \lambda_i$, $i = 1, \ldots, n$. Then (by E1.1.6 and E1.1.7) the eigenvalues of $(A - \bar{\lambda} I)^{-1}$ are $(\lambda_i - \bar{\lambda})^{-1}$, $i = 1, \ldots, n$, corresponding to the eigenvectors $\mathbf{u}_1, \ldots, \mathbf{u}_n$. Hence

$$\mathbf{x} = (A - \bar{\lambda} I)^{-1} \mathbf{b} = \sum_{i=1}^{n} \frac{\alpha_i}{\lambda_i - \bar{\lambda}} \mathbf{u}_i. \tag{6}$$

To understand the utility of this basic representation, suppose, for example, that $\alpha_1 \doteq \cdots \doteq \alpha_n \doteq 1$, so that $\mathbf{b}$ has "equal weight" in all the eigendirections, that $\bar{\lambda}$ is a "good" approximation to λ_1, say

$$|\lambda_1 - \bar{\lambda}| \doteq 10^{-8},$$

that λ_1 is well-separated from the other eigenvalues, say $|\lambda_1 - \lambda_i| \ge 10^{-2}$,

$i = 2, \ldots, n$, and that the eigenvectors have been normalized so that $\|\mathbf{u}_i\| = 1$, $i = 1, \ldots, n$. Then (6) yields

$$\left\| \frac{\pm \mathbf{x}}{\|\mathbf{x}\|} - \mathbf{u}_1 \right\| \leq 3(n - 1)10^{-6}$$

where the sign is chosen so that $\pm \mathbf{x}$ and $\mathbf{u}_1$ point in the same direction. This estimate implies that $\pm \mathbf{x}/\|\mathbf{x}\|$ is an approximation to $\mathbf{u}_1$ good to approximately six decimal places (for small n).

In general, the goodness of the approximation (6) to $\mathbf{u}_1$ depends on three factors:

(a) $|\lambda_1 - \bar{\lambda}|$ should be as small as possible;

(b) $|\lambda_i - \bar{\lambda}|$, $i = 2, \ldots, n$ should be as large as possible;

(c) $|\alpha_1|/|\alpha_i|$, $i = 2, \ldots, n$ should be as large as possible.

Condition (c) simply states that the right-hand side, $\mathbf{b}$, should be as much in the direction of the sought eigenvector as possible; if $\mathbf{b}$ is in the direction $\mathbf{u}_1$, then so is $\mathbf{x}$, while if $\mathbf{b}$ contains no component of $\mathbf{u}_1$ (i.e., if $\alpha_1 = 0$), then neither does $\mathbf{x}$. Condition (b) indicates that even if (a) holds, it may be that λ_1 is in a "cluster" of eigenvalues and $\mathbf{x}$ will be contaminated by contributions from the corresponding eigenvectors. Condition (a), in conjunction with (b), implies that we wish the coefficient matrix $A - \bar{\lambda}I$ to be as ill conditioned as possible; this follows from (2.1.10). This ill conditioning is reflected in the fact that the rounding error produced in solving (5) may, in fact, cause a large error in $\mathbf{x}$; but if (b) holds, the effect is that $\mathbf{x}$ will be in error primarily in the direction $\mathbf{u}_1$, and it is precisely this direction we are attempting to find. Hence, this ill conditioning does not cause a deleterious effect.

The above analysis has straightforward implications for a posteriori bounds based on the residual. Suppose that A has a small eigenvalue and that $\mathbf{y}$ is a corresponding eigenvector. Then if $\bar{\mathbf{x}}$ is an approximate solution of $A\mathbf{x} = \mathbf{b}$, it follows that

$$\mathbf{r} = A\bar{\mathbf{x}} - \mathbf{b} \doteq A(\bar{\mathbf{x}} + \mathbf{y}) - \mathbf{b}. \tag{7}$$

For example, suppose that the small eigenvalue is $\lambda \doteq 10^{-8}$. Then (7) shows that the effect on the residual of adding $\mathbf{y}$ to $\bar{\mathbf{x}}$ is 10^{-8} of the effect on the approximate solution itself. The implication of this is that large errors in the solution in the direction of $\mathbf{y}$ are not easily detectable by examining only the residual vector.

EXERCISES

E2.2.1 Consider the system†

$$0.780x_1 + 0.563x_2 = 0.217$$
$$0.913x_1 + 0.659x_2 = 0.254.$$

Compute the residual vector, $A\mathbf{x} - \mathbf{b}$, for the two approximate solutions $(0.341, -0.087)$ and $(0.999, -1.001)$ and "conclude" from the size of these residuals which is the better solution. Then show that $(1, -1)$ is the exact solution.

E2.2.2 Let‡

$$A = \begin{bmatrix} 9999. & 9998. \\ 10000. & 9999. \end{bmatrix}, \qquad B = \begin{bmatrix} 9999.9999 & -9997.0001 \\ -10001. & 9998. \end{bmatrix}.$$

Show that

$$BA - I = \begin{bmatrix} 19998. & 19995. \\ -19999. & -19996. \end{bmatrix}, \qquad AB - I = \begin{bmatrix} 0.0001 & 0.0001 \\ 0 & 0 \end{bmatrix}.$$

Using these residuals, compute a bound for $A^{-1} - B$.

E2.2.3 Show that the matrix

$$A = \begin{bmatrix} 2 & 1 \\ 1 & 2 \end{bmatrix}$$

has eigenvalues 1 and 3 and compute the eigenvector $\mathbf{u}$ corresponding to 3. Now suppose that $\lambda = 2.99$ is an approximate eigenvalue and compute a corresponding eigenvector by solving the system

$$(A - \lambda I)\mathbf{x} = (1, 2)^{\mathrm{T}}.$$

Normalize both $\mathbf{u}$ and $\mathbf{x}$ in some norm and comment on the accuracy of $\mathbf{x}$.

READING

Excellent references for the material of this chapter are Wilkinson [1965], Wilkinson [1963], and Forsythe and Moler [1967]. See also Householder [1964] for a more theoretical treatment.

† This example is due to C. B. Moler.
‡ This example is due to G. E. Forsythe, *Amer. Math. Monthly* **77** (1970), 931–955.

CHAPTER 3

EIGENVALUES AND EIGENVECTORS

3.1 CONTINUITY RESULTS

In this chapter we shall consider the following question. How do the roots of a polynomial

$$p(\lambda) = \lambda^n + a_{n-1}\lambda^{n-1} + \cdots + a_1\lambda + a_0 \tag{1}$$

vary as functions of the coefficients $a_0, \ldots, a_{n-1}$, and, more generally, how do the eigenvalues and eigenvectors of a matrix A vary as functions of the elements a_{ij} of A? Since the roots of (1) are just the eigenvalues of the **companion matrix**

$$\begin{bmatrix} -a_{n-1} & \cdots & -a_0 \\ 1. & 0 & \cdots & 0 \\ & & \ddots & \vdots \\ O & & 1 & 0 \end{bmatrix} \tag{2}$$

of p (see E3.1.1), the first question is really a special case of the second.

We begin with a basic theorem that the roots of a polynomial are continuous functions of its coefficients. The proof will be based on the following famous result in complex variable theory. (For a proof, see essentially any intermediate-level book on complex variables.)

Rouche's Theorem Assume that f and g are analytic and single valued in a domain D and on its boundary $\dot{D}$, and that on $\dot{D}$, $|f(z) - g(z)| < |f(z)|$. Then f and g have exactly the same number of zeros in D.

3.1.1 (Continuity of Roots Theorem) Let p be the polynomial (1) with real or complex coefficients $a_0, \ldots, a_{n-1}$ and with roots $\lambda_1, \ldots, \lambda_n$. Then for any sufficiently small $\varepsilon > 0$, there is a $\delta > 0$ such that the roots μ_i of any polynomial

$$q(\lambda) = \lambda^n + b_{n-1}\lambda^{n-1} + \cdots + b_0 \tag{3}$$

with $|b_i - a_i| \leq \delta, i = 0, \ldots, n-1$, can be ordered so that

$$|\lambda_i - \mu_i| \leq \varepsilon, \qquad i = 1, \ldots, n. \tag{4}$$

Proof†: Let $\gamma_1, \ldots, \gamma_m$ be the distinct roots of p, assume that $0 < \varepsilon < \frac{1}{2}|\gamma_i - \gamma_j|$ for $i, j = 1, \ldots, m$, $i \neq j$ (if $m > 1$), and define the disks $D_i = \{z : |z - \gamma_i| \leq \varepsilon\}$ with boundaries $\dot{D}_i, i = 1, \ldots, m$. Then, since the D_i are disjoint, p does not vanish on any $\dot{D}_i$ and, because of the continuity of p and the compactness of $\dot{D}_i$, there is an $m_i > 0$ so that

$$|p(z)| \geq m_i, \qquad z \in \dot{D}_i.$$

Next let

$$M_i = \max\{|z^{n-1}| + \cdots + |z| + 1 : z \in \dot{D}_i\}.$$

Then

$$|q(z) - p(z)| \leq \delta M_i, \qquad z \in \dot{D}_i$$

and if δ is chosen sufficiently small that $M_i\delta < m_i$ it follows that

$$|p(z)| > |q(z) - p(z)|, \qquad z \in \dot{D}_i.$$

Hence, Rouche's theorem ensures that p and q have precisely the same number of roots in each D_i, and the result follows. $\$\$\$$

It is important to note that Theorem 3.1.1 does *not* imply that the roots of a polynomial are well conditioned with respect to changes in its coefficients. Consider, for example, the polynomial

$$\lambda^n = 0 \tag{5}$$

---

† This proof is adapted from that given in J. Franklin, "Matrix Theory," Prentice-Hall, Englewood Cliffs, New Jersey, 1968.

that is, the coefficients $a_{n-1}, \ldots, a_0$ of (1) are all zero. Now change the coefficient a_0 by ε and consider the polynomial

$$\lambda^n - \varepsilon = 0 \qquad (6)$$

where we assume ε to be positive. Then the roots of (6) are $\lambda_j = \omega^j \varepsilon^{1/n}$, $j = 1, \ldots, n$, where $\omega = \exp(i2\pi/n)$ is an nth root of unity. Hence a change of ε in one coefficient produces a change of modulus $\varepsilon^{1/n}$ in the roots.

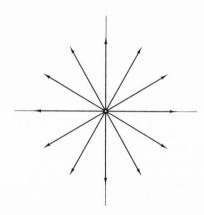

Figure 3.1.1

Note also that the multiple root $\lambda = 0$ has "scattered" into n simple roots symmetrically placed about the origin in the complex plane (see Figure 3.1.1). Now suppose that $n = 100$ and $\varepsilon = 10^{-100}$. Then $\varepsilon^{1/n} = 10^{-1}$ so that the magnitude of the change in the root is 10^{99} times the change in the coefficient. This is, of course, an extreme example but deleterious effects can also occur for polynomials of very low degree. Consider, for example, the quadratic

$$\lambda^2 - 2\lambda + 0.99999999$$

with roots 1 ± 10^{-4}. If we change the constant term to 1, then the roots both become 1, so that a change of 10^{-8} in one coefficient has caused changes 10^4 times as large in the roots.

The above discussion is indicative of the fact that multiple roots of a polynomial are ill conditioned. But "well-separated" roots may also be ill conditioned as the following example shows.† Consider the polynomial

$$p(\lambda) = \prod_{k=1}^{20} (\lambda - k) \tag{7}$$

and the perturbed polynomial $\hat{p}(\lambda) = p(\lambda) - \varepsilon\lambda^{19}$. For $\varepsilon = 2^{-23} \doteq 10^{-7}$ the roots are, rounded to the number of places given,

1.0	6.0	$11.8 \pm 1.7i$
2.0	7.0	$14.0 \pm 2.5i$
3.0	8.0	$16.7 \pm 2.8i$
4.0	8.9	$19.5 \pm 1.9i$
5.0	$10.1 \pm 0.6i$	20.8

Since the coefficient of x^{19} in $p(x)$ is 210, we see that a change in this one coefficient of about $10^{-7}\%$ has produced such large changes in some of the roots that they have become complex.

We turn next to the effect of changes in the elements of a matrix upon the eigenvalues of the matrix. A first basic result is the following.

3.1.2 (Continuity of Eigenvalues) The eigenvalues of a matrix are continuous functions of the elements of the matrix.

This is an immediate consequence of Theorem 3.1.1 together with the fact (E3.1.4) that the coefficients of the characteristic equation are continuous functions of the elements of the matrix.

It is important to note that **3.1.2** does *not*, in general, hold for the eigenvectors of a matrix (E3.1.5). However, it does hold for simple eigenvalues as the following result shows.

3.1.3 (Continuity of Eigenvectors) Let λ be a simple eigenvalue of $A \in L(C^n)$ and let $\mathbf{x} \neq \mathbf{0}$ be a corresponding eigenvector. Then for $E \in L(C^n)$, $A + E$ has an eigenvalue $\lambda(E)$ and eigenvector $\mathbf{x}(E)$ such that

$$\lambda(E) \to \lambda \quad \text{and} \quad \mathbf{x}(E) \to \mathbf{x} \quad \text{as} \quad E \to 0.$$

† Whether the roots of (7) are, in fact, well separated is open to interpretation. The coefficients a_j of (7) range in magnitude up to 10^{20} and the l_1 norm of the companion matrix (2) is about 10^{21}. Relative to this norm, the roots of (7) are almost multiple. See Wilkinson [1963, p. 43].

Proof: Since λ is simple, it follows immediately by consideration of the Jordan form of A that $A - \lambda I$ has rank $n - 1$. Hence, there are indices i and j such that $\mathbf{x}$ satisfies the system

$$\sum_{m \neq j} (a_{km} - \delta_{km} \lambda) x_m = (a_{kj} - \delta_{kj} \lambda) x_j, \qquad k \neq i \qquad (8)$$

where the coefficient matrix of this system is nonsingular, and where δ_{km} is the Kronecker delta. Without loss of generality, we may assume that $x_j = 1$.

Now let $\lambda(E)$ be an eigenvalue of $A + E$ such that $\lambda(E) \to \lambda$ as $E \to 0$; such a sequence of eigenvalues exists by 3.1.2. Moreover, for sufficiently small $\|E\|$, $\lambda(E)$ is also simple and the matrix $A + E - \lambda(E)I$ is also of rank $n - 1$; in fact, it follows from Theorem 2.1.1 that the submatrix obtained by deleting the ith row and jth column is nonsingular. Therefore, the system

$$\sum_{m \neq j} \left(a_{km} + e_{km} - \delta_{km} \lambda(E) \right) x_m(E) = \left(a_{kj} + e_{kj} - \delta_{kj} \lambda(E) \right), \qquad k \neq i$$

has a unique solution $x_m(E)$, $m \neq j$, which, again by 2.1.1, is a continuous function of E. That is, $x_m(E) \to x_m$ as $E \to 0$ for $m \neq j$, and we may choose $x_j(E) = x_j = 1$. \$\$\$

EXERCISES

E3.1.1   If A is the matrix (1), show that

$$\det(\lambda I - A) = \lambda^n + a_{n-1}\lambda^{n-1} + \cdots + a_0.$$

E3.1.2   Denote the roots of the quadratic equation $x^2 + bx + c = 0$ by $x_{\pm} = x_{\pm}(b, c)$, considered as functions of b and c. Show that

$$\frac{\partial x_{\pm}}{\partial b} = -\frac{1}{2} \pm \frac{1}{2} \frac{b}{(b^2 - 4c)^{1/2}}, \qquad \frac{\partial x_{\pm}}{\partial c} = \mp \frac{1}{(b^2 - 4c)^{1/2}}$$

and that, therefore, these derivatives are infinite at a multiple root. Comment on the relationship of this result to the conditioning of the roots.

E3.1.3   Consider the $n \times n$ matrices†

† In this context, B is sometimes called the **Forsythe matrix**.

$$A = \begin{bmatrix} a & 1 & & \text{O} \\ & a & \ddots & \\ & & \ddots & 1 \\ \text{O} & & & a \end{bmatrix}, \qquad B = \begin{bmatrix} a & 1 & & \text{O} \\ & \cdot & \ddots & \\ & & \cdot & 1 \\ \varepsilon & & \cdot & a \end{bmatrix},$$

where A and B differ only in the $(n, 1)$ element. Show that the eigenvalues λ_i of B satisfy

$$|\lambda_i - a| = |\varepsilon|^{1/n}, \qquad i = 1, \ldots, n.$$

Consider the special case $\varepsilon = 10^{-n}$ and discuss the percentage changes in the eigenvalues as a function of both n and a.

E3.1.4 Let $A \in L(C^n)$. Show that det A is a continuous function of the elements of A and, more generally, that the coefficients of the characteristic polynomial $\det(A - \lambda I) = 0$ are continuous functions of the elements of A.

E3.1.5† Show that the matrix

$$A(\varepsilon) = \begin{bmatrix} 1 + \varepsilon \cos(2/\varepsilon) & -\varepsilon \sin(2/\varepsilon) \\ -\varepsilon \sin(2/\varepsilon) & 1 - \varepsilon \cos(2/\varepsilon) \end{bmatrix}, \qquad \varepsilon \neq 0$$

has eigenvalues $1 \pm \varepsilon$ and corresponding eigenvectors $(\sin(1/\varepsilon), \cos(1/\varepsilon))^{\mathrm{T}}$, $(\cos(1/\varepsilon), -\sin(1/\varepsilon))^{\mathrm{T}}$. Conclude that these eigenvectors do not tend to a limit as $\varepsilon \to 0$ even though $\lim_{\varepsilon \to 0} A(\varepsilon)$ exists.

3.2 THE GERSCHGORIN AND BAUER–FIKE THEOREMS

Although the continuity results of the previous section are extremely important, they do not give any information as to the magnitude of the changes in the eigenvalues or eigenvectors as a result of changes in the elements of the matrix. We now give several such results. The first is an example of a **localization theorem**, that is, it gives regions in the complex plane which are known to contain eigenvalues of the matrix.

3.2.1 (Gerschgorin Circle Theorem) For $A = (a_{ij}) \in L(C^n)$, define the disks

$$R_i = \left\{ z : |a_{ii} - z| \leq \sum_{j \neq i} |a_{ij}| \right\}, \qquad i = 1, \ldots, n$$

† This example is due to J. W. Givens.

in the complex plane. Then every eigenvalue of A lies in the union $S = \bigcup_{i=1}^{n} R_i$. Moreover, if $\hat{S}$ is a union of m disks R_i such that $\hat{S}$ is disjoint from all other disks, then $\hat{S}$ contains precisely m eigenvalues (counting multiplicities) of A.

Proof: Suppose that λ is an eigenvalue of A with corresponding eigenvector $\mathbf{x} \neq 0$ and let i be an index such that

$$|x_i| = \max_{1 \le j \le n} |x_j|.$$

Clearly, $x_i \neq 0$ and hence it follows from

$$|a_{ii} - \lambda|\,|x_i| = |a_{ii}x_i - \lambda x_i| = \left| \sum_{j \neq i} a_{ij} x_j \right| \le \sum_{j \neq i} |a_{ij}|\,|x_i|$$

that $\lambda \in R_i$. For the second part, define the family of matrices $A_t = D + tB$ for $t \in [0, 1]$, where $D = \mathrm{diag}(a_{11}, \ldots, a_{nn})$ and $B = A - D$. Let $R_i(t)$ denote the disk with center a_{ii} and radius $t \sum_{j \neq i} |a_{ij}|$. Without loss of generality we may assume that $\hat{S} = \bigcup_{i=1}^{m} R_i$. Set $\hat{S}(t) = \bigcup_{i=1}^{m} R_i(t)$ and $\tilde{S}(t) = \bigcup_{i=m+1}^{n} R_i(t)$. By assumption, $\tilde{S}(1)$ is disjoint from $\hat{S}$ and, clearly, this implies that $\tilde{S}(t)$ and $\hat{S}(t)$ are disjoint for all $t \in [0, 1]$. In particular, $\hat{S}(0)$ must contain precisely m eigenvalues of A_0, namely, $a_{11}, \ldots, a_{mm}$. Now the first part of the proof implies that the eigenvalues of A_t are all contained in $\hat{S}(t) \cup S(t)$ for all $t \in [0, 1]$. But since $\hat{S}(t)$ and $\tilde{S}(t)$ are disjoint, the continuity theorem 3.1.1 shows that an eigenvalue of A_t cannot "jump" from $\hat{S}(t)$ to $\tilde{S}(t)$ (or conversely) as t increases. Hence, since $\hat{S}(0)$ contains precisely m eigenvalues of A_0, $\hat{S}(t)$ must contain precisely m eigenvalues of A_t for all $t \in [0, 1]$. The situation is depicted in Figure 3.2.1. $\$\$\$$

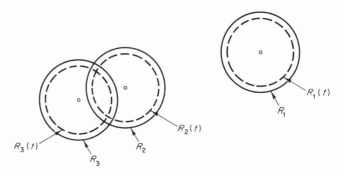

Figure 3.2.1

As a very simple example of the use of **3.2.1** consider the matrix

$$A = \begin{bmatrix} 2 & 2 & 2 \\ 2 & 4 & 1 \\ 1 & 1 & 10 \end{bmatrix}.$$

From $\rho(A) \le \|A\|_\infty$ we obtain the very crude estimate that all eigenvalues lie in the disk $\{z : |z| \le 12\}$, but by **3.2.1** we have the sharper result that the eigenvalues lie in

$$\{z : |z - 2| \le 4\} \cup \{z : |z - 4| \le 3\} \cup \{z : |z - 10| \le 2\}.$$

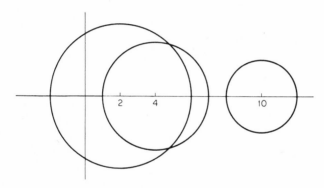

Figure 3.2.2

Moreover, since the last disk is isolated, **3.2.1** ensures that it contains precisely one eigenvalue. The estimates are depicted in Figure 3.2.2. A more interesting example is the following†:

If the off-diagonal elements of a matrix are very small, then we expect the eigenvalues to be approximately equal to the diagonal elements. This statement can be made more precise by Gerschgorin's theorem. Consider the matrix

$$B = \begin{bmatrix} 0.9 & & \bigcirc \\ & 0.4 & \\ \bigcirc & & 0.2 \end{bmatrix} + 10^{-5} \begin{bmatrix} 0.1 & 0.4 & -0.2 \\ -0.1 & 0.5 & 0.1 \\ 0.2 & 0.1 & 0.3 \end{bmatrix}. \tag{1}$$

† Given in Wilkinson [1965, p. 74].

Then by **3.2.1** the exact eigenvalues of B satisfy

$$|\lambda_1 - (0.9 + 10^{-6})| \quad \le 0.6 \cdot 10^{-5} \quad \text{or} \quad |\lambda_1 - 0.9| \le 0.7 \cdot 10^{-5}$$
$$|\lambda_2 - (0.4 + 0.5 \cdot 10^{-5})| \le 0.2 \cdot 10^{-5} \quad \text{or} \quad |\lambda_2 - 0.4| \le 0.7 \cdot 10^{-5} \quad (2)$$
$$|\lambda_3 - (0.2 + 0.3 \cdot 10^{-5})| \le 0.3 \cdot 10^{-5} \quad \text{or} \quad |\lambda_3 - 0.2| \le 0.6 \cdot 10^{-5}$$

While it would be rare that a matrix such as B would be presented to us a priori, it might well arise by an a posteriori analysis of a computed eigensystem. That is, suppose that for some 3×3 matrix A we have computed approximate eigenvalues 0.9, 0.4, and 0.2, and corresponding approximate eigenvectors which we take to be the columns of a matrix P. Now assume that the matrix B of (1) is obtained by $B = P^{-1}AP$, where we suppose, for simplicity, that this similarity transformation is done exactly. Then B is similar to A and has the same eigenvalues, so that the above procedure provides an a posteriori error analysis.

Note that while the second set of inequalities in (2) gives bounds on the originally computed eigenvalues, the first set gives slightly improved bounds for a corrected set of eigenvalues obtained by adding in the diagonal elements of the second matrix of (1). A more interesting correction procedure is to perform diagonal similarity transformations on B. Multiply the first row of B by 10^{-5} and the first column by 10^5. This is a similarity transformation with the diagonal matrix $D_1 = \text{diag}(10^{-5}, 1, 1)$ and hence

$$D_1 B D_1^{-1} = \begin{bmatrix} 0.9 & & \bigcirc \\ & 0.4 & \\ \bigcirc & & 0.2 \end{bmatrix} + \begin{bmatrix} 0.1 \cdot 10^{-5} & 0.4 \cdot 10^{-10} & -0.2 \cdot 10^{-10} \\ -0.1 & 0.5 \cdot 10^{-5} & 0.1 \cdot 10^{-5} \\ 0.2 & 0.1 \cdot 10^{-5} & 0.3 \cdot 10^{-5} \end{bmatrix}.$$

Since the first disk still does not overlap the other two, Theorem **3.2.1** now ensures that

$$|\lambda_1 - (0.9 + 10^{-6})| \le 6 \cdot 10^{-11}.$$

Analogous similarity transformations using $D_2 = \text{diag}(1, 10^{-5}, 1)$ and $D_3 = \text{diag}(1, 1, 10^{-5})$ yield

$$|\lambda_2 - (0.4 + 0.5 \cdot 10^{-5})| \le 2 \cdot 10^{-11}, \qquad |\lambda_3 - (0.2 + 10^{-6})| \le 5 \cdot 10^{-11}.$$

Clearly these bounds are rather spectacularly better than those of (2) and at the cost of very little effort.

We summarize the above correction procedure in the following statement.

3.2.2   (**Wilkinson Correction Procedure**)   For $A \in L(C^n)$ assume that the disk

$$R = \left\{ z : |z - a_{ii}| \leq \alpha \sum_{j \neq i} |a_{ij}| \right\}$$

is disjoint from all the disks

$$\left\{ z : |z - a_{kk}| \leq \alpha^{-1} |a_{ki}| + \sum_{j \neq k, i} |a_{kj}| \right\}, \qquad k = 1, \dots, n, \quad k \neq i.$$

Then R contains precisely one eigenvalue of A.

We turn our attention next to perturbation theorems, that is, to theorems which give an estimate of how much the eigenvalues of $A + E$ can differ from those of A. The first result, a special case of which has already been used in the previous example, is an easy consequence of **3.2.1**.

3.2.3   Assume that $A = PDP^{-1} \in L(C^n)$, where $D = \text{diag}(\lambda_1, \dots, \lambda_n)$, and that $A + E$ has eigenvalues $\mu_1, \dots, \mu_n$. Then given any μ_j there is a λ_i such that

$$|\lambda_i - \mu_j| \leq \|P^{-1}EP\|_\infty. \tag{3}$$

Moreover, if λ_i is an eigenvalue of multiplicity m and the disk

$$R = \{ z : |z - \lambda_i| \leq \|P^{-1}EP\|_\infty \} \tag{4}$$

is disjoint from the disks

$$\{ z : |z - \lambda_k| \leq \|P^{-1}EP\|_\infty \}, \qquad \lambda_k \neq \lambda_i \tag{5}$$

then R contains precisely m eigenvalues of $A + E$.

Proof: Set $C = P^{-1}(A + E)P$; then C has eigenvalues $\mu_1, \dots, \mu_n$. Denote the elements of $P^{-1}EP$ by b_{ij}, $i, j = 1, \dots, n$. Then the diagonal elements of C are $\lambda_k + b_{kk}$ and by **3.2.1** there is an i such that

$$|\lambda_i + b_{ii} - \mu_j| \leq \sum_{k \neq i} |b_{ik}|.$$

Hence

$$|\lambda_i - \mu_j| \leq \sum_{k=1}^{n} |b_{ik}| \leq \|P^{-1}EP\|_\infty. \tag{6}$$

The last statement follows directly from the last statement of **3.2.1**. $$$

Theorem 3.2.3 was proved in the l_∞ norm, and it also holds, trivially, in the l_1 norm (E3.2.8). For theoretical purposes, we would like the result to be true in the l_2 norm. It turns out that this is the case and that, indeed, it holds for any norm with the following property.

3.2.4 Definition A norm $\|\cdot\|$ on R^n or C^n is **monotonic** (or **absolute**) if for any diagonal matrix $D = \text{diag}(d_1, \ldots, d_n)$ the corresponding operator norm satisfies

$$\|D\| = \max_{i=1,\ldots,n} |d_i|.$$

It follows immediately from 1.3.4 and 1.3.5 that the l_1, l_2, and l_∞ norms are all monotonic. Indeed, any l_p norm is monotonic as an immediate consequence of the following characterization (which is usually given as the definition). We note, however, that elliptic norms are not monotonic in general (E3.2.9).

3.2.5† A norm on R^n (or C^n) is monotonic if and only if $|x_i| \leq |y_i|$, $i = 1, \ldots, n$ implies that $\|\mathbf{x}\| \leq \|\mathbf{y}\|$.

We can now prove the desired extension of 3.2.3.

3.2.6 (Bauer–Fike Theorem)‡ The conclusions of 3.2.3 hold in any monotonic norm.

Proof: Again, let $C = P^{-1}(A + E)P = D + B$ with $B = P^{-1}EP$, and consider any eigenvalue μ of C. If $D - \mu I$ is singular, then $\mu = \lambda_i$ for some i, and the theorem is proved. Hence, assume that $D - \mu I$ is not singular. Then

$$C - \mu I = (D - \mu I)[I + (D - \mu I)^{-1}B],$$

and since $C - \mu I$ is singular, it follows that $I + (D - \mu I)^{-1}B$ is singular. Therefore, $(D - \mu I)^{-1}B$ has -1 as an eigenvalue so that

$$\|(D - \mu I)^{-1}\| \, \|B\| \geq \|(D - \mu I)^{-1}B\| \geq 1. \tag{7}$$

---

† For a proof, see Householder [1964, p. 47].

‡ F. Bauer and C. Fike, *Numer. Math.* **2** (1960), 137–141. Actually a slightly sharper result [see J. Osborn, *Numer. Math.* **13** (1969), 152–153] already had been proved by J. T. Schwartz, *Pac. J. Math.* **4** (1954), 415–458, in a different context.

Since the norm is monotonic

$$\|(D - \mu I)^{-1}\| = \max_i \frac{1}{|\lambda_i - \mu|} = \frac{1}{\min_i |\lambda_i - \mu|},$$

and, hence, by (7),

$$\min_i |\lambda_i - \mu| \leq \|P^{-1}EP\|.$$

The second part follows precisely as in the Gerschgorin theorem 3.2.1. $$$

We have given an example of an a posteriori analysis based on Gerschgorin's theorem, and under the assumption that a complete approximate eigensystem was known. We next give a somewhat different result.

Suppose that λ and $\mathbf{x}$ are an approximate eigenvalue and eigenvector of $A \in L(C^n)$. To test the accuracy of these approximations it is natural to form the residual vector $\mathbf{r} = A\mathbf{x} - \lambda\mathbf{x}$. If $\mathbf{r} = 0$, then, of course, λ and $\mathbf{x}$ are exact, but if $\|\mathbf{r}\|$ is only small, then λ may be far from an eigenvalue (E3.2.10). The best estimates we can obtain depend upon the conditioning of the eigenvalue.

3.2.7   Assume that $A = PDP^{-1} \in L(C^n)$ with $D = \operatorname{diag}(\lambda_1, \ldots, \lambda_n)$. Then, in any monotonic norm, if $\|A\mathbf{x} - \lambda\mathbf{x}\| \leq \varepsilon$ with $\|\mathbf{x}\| = 1$ we have

$$\min_{1 \leq i \leq n} |\lambda_i - \lambda| \leq \varepsilon \|P^{-1}\| \, \|P\|. \tag{8}$$

Proof: We may assume that $D - \lambda I$ is nonsingular or else (8) is trivial. Then

$$\mathbf{r} = A\mathbf{x} - \lambda\mathbf{x} = P(D - \lambda I)P^{-1}\mathbf{x}$$

so that

$$1 = \|\mathbf{x}\| = \|P(D - \lambda I)^{-1}P^{-1}\mathbf{r}\| \leq \varepsilon \|P\| \, \|P^{-1}\|/\min_i |\lambda_i - \lambda|$$

which is (8). $$$

Theorems **3.2.3** and **3.2.6** show that a crucial factor in the analysis of the eigenvalues of $A + E$ is the matrix $P^{-1}EP$. We now make a somewhat finer analysis, still under the assumption that

$$P^{-1}AP = D = \operatorname{diag}(\lambda_1, \ldots, \lambda_n).$$

Let $\mathbf{x}_1, \ldots, \mathbf{x}_n$ denote the columns of P and $s_1^{-1}\mathbf{y}_1^{\mathrm{T}}, \ldots, s_n^{-1}\mathbf{y}_n^{\mathrm{T}}$ the rows of P^{-1}, where we assume that $\|\mathbf{x}_i\|_2 = \|\mathbf{y}_i\|_2 = 1$, $i = 1, \ldots, n$. From the fact that $P^{-1}P = I$, it follows that $\mathbf{y}_i^{\mathrm{T}}\mathbf{x}_j = s_i \delta_{ij}$. Suppose, now, that λ_i is a simple eigenvalue. Then for sufficiently small $\|E\|_\infty$, the disk R of (4) will be isolated from the disks (5) and 3.2.3 ensures that R contains precisely one eigenvalue μ_j of $A + E$. Moreover, from (6) we have

$$|\lambda_i - \mu_j| \le \sum_{k=1}^n |\mathbf{y}_i^{\mathrm{T}} E \mathbf{x}_k| / |s_i|.$$

The crucial factor in this estimate is the quantity $s_i = \mathbf{y}_i^{\mathrm{T}}\mathbf{x}_i$, and we will use its reciprocal as a condition number of the eigenvalue λ_i.

3.2.8   Definition   Let λ be a simple eigenvalue of $A \in L(C^n)$ and let $A\mathbf{x} = \lambda\mathbf{x}$ and $\mathbf{y}^{\mathrm{T}}A = \lambda\mathbf{y}^{\mathrm{T}}$, where $\mathbf{x}, \mathbf{y} \ne \mathbf{0}$. Then $\mathbf{y}$ is a **left eigenvector** of A (and in this context $\mathbf{x}$ is a **right eigenvector**). Assume that $\|\mathbf{x}\|_2 = \|\mathbf{y}\|_2 = 1$. Then $|\mathbf{y}^{\mathrm{T}}\mathbf{x}|^{-1}$ is the **condition number** of λ.

Since we may take $\mathbf{x}$ as the ith column of P and a scalar multiple of $\mathbf{y}^{\mathrm{T}}$ as the ith row of P^{-1}, where P is the matrix such that $P^{-1}AP$ is the Jordan form of A with λ in the ith diagonal position, it follows from $P^{-1}P = I$ that $\mathbf{y}^{\mathrm{T}}\mathbf{x} \ne 0$. Note that it is $\mathbf{y}^{\mathrm{T}}$ and not $\mathbf{y}^{\mathrm{H}}$ which is used, even if $\mathbf{y}$ is complex. Note also that the condition number is bounded by $K_2(P)$ (see E3.2.6).

In order to substantiate somewhat further the role that $|\mathbf{y}^{\mathrm{T}}\mathbf{x}|^{-1}$ plays as a condition number, we end this section with the limiting case of 3.2.3 as $E \to 0$.

3.2.9   Assume that $A \in L(C^n)$ is similar to a diagonal matrix and let λ be a simple eigenvalue with left and right eigenvectors $\mathbf{y}$ and $\mathbf{x}$. Let $C \in L(C^n)$ be fixed but arbitrary and set $E = \varepsilon C$. Assume that ε is so small that the disk R of (4) is isolated from the disks (5) and let $\lambda(\varepsilon)$ be the unique eigenvalue of $A + E$ in R. Then

$$\lambda'(0) = \lim_{\varepsilon \to 0} \frac{\lambda(\varepsilon) - \lambda}{\varepsilon} = \frac{\mathbf{y}^{\mathrm{T}} C \mathbf{x}}{\mathbf{y}^{\mathrm{T}} \mathbf{x}}.$$

Proof: Let $\mathbf{x}(\varepsilon)$ be an eigenvector corresponding to $\lambda(\varepsilon)$ so that

$$(A + \varepsilon C)\mathbf{x}(\varepsilon) = \lambda(\varepsilon)\mathbf{x}(\varepsilon).$$

Then subtract $A\mathbf{x} = \lambda\mathbf{x}$ and add and subtract $\lambda\mathbf{x}(\varepsilon)$ to obtain

$$A(\mathbf{x}(\varepsilon) - \mathbf{x}) + \varepsilon C\mathbf{x}(\varepsilon) = (\lambda(\varepsilon) - \lambda)\mathbf{x}(\varepsilon) + \lambda(\mathbf{x}(\varepsilon) - \mathbf{x}).$$

Now multiply through by $\mathbf{y}^T$. Then the first and last terms cancel to leave

$$\varepsilon \mathbf{y}^T C\mathbf{x}(\varepsilon) = (\lambda(\varepsilon) - \lambda)\mathbf{y}^T\mathbf{x}(\varepsilon).$$

By 3.1.3, it follows that we may assume that $\mathbf{x}(\varepsilon)$ is chosen so that $\mathbf{x}(\varepsilon) \to \mathbf{x}$ as $\varepsilon \to 0$. $\$\$\$$

EXERCISES

E3.2.1   Show that Theorem 3.2.1 remains valid if R_i is replaced by $\{z : |a_{ii} - z| \le \sum_{j \ne i} |a_{ji}|\}$.

E3.2.2   Suppose that

$$A = \begin{bmatrix} 0.1 & 0 & 1.0 & 0.2 \\ 0.2 & 0.3 & 0.1 & 0.3 \\ 0.1 & 0.1 & 0.1 & 0.2 \\ 0.1 & 0.2 & 0.2 & 0.1 \end{bmatrix}$$

and observe that $\|A\|_i > 1$, $i = 1, \infty$. Show, however, that $\rho(A) < 1$ by considering DAD^{-1} for a suitable diagonal matrix D.

E3.2.3   Consider the matrix

$$A = \begin{bmatrix} 1 & 10^{-5} & 10^{-5} \\ 10^{-5} & 2 & 10^{-5} \\ 10^{-5} & 10^{-5} & 3 \end{bmatrix}$$

with eigenvalues $\lambda_1 \le \lambda_2 \le \lambda_3$. Give the best bounds you can for $|\lambda_j - j|$, $j = 1, 2, 3$, by means of 3.2.1 and 3.2.2.

E3.2.4   Give a formal proof of 3.2.2.

E3.2.5   Give an example of matrices A and E such that $\|E\|$ is arbitrarily large but A and $A + E$ have the same eigenvalues.

E3.2.6   Let $A \in L(C^n)$ have distinct eigenvalues, and let $D = P^{-1}AP$ be diagonal, where it is assumed that the columns of P are normalized to euclidean length 1. Show that

$$\max_{1 \le i \le n} c_i \le K_2(P)$$

where $c_1, \ldots, c_n$ are the condition numbers of the eigenvalues.

E3.2.7 Suppose that $A \in L(R^n)$ has eigenvalues $\lambda_1, \ldots, \lambda_n$, corresponding linearly independent right eigenvectors $\mathbf{u}_1, \ldots, \mathbf{u}_n$, and left eigenvectors $\mathbf{v}_1, \ldots, \mathbf{v}_n$, normalized so that $\|\mathbf{u}_i\|_2 = \|\mathbf{v}_i\|_2 = 1$, $i = 1, \ldots, n$. Set $c_i = |\mathbf{v}_i^T\mathbf{u}_i|^{-1}$, $i = 1, \ldots, n$. Show that the solution of $A\mathbf{x} = \mathbf{b}$ satisfies

$$\|\mathbf{x}\|_2 \leq \|\mathbf{b}\|_2 \sum_{i=1}^{n} \frac{c_i}{|\lambda_i|}.$$

Conclude that A is ill conditioned (with respect to the inverse problem) only if either some λ_i is small or some c_i is large.

E3.2.8 Use E3.2.1 to conclude that 3.2.3 holds in the l_1 norm.

E3.2.9 Let $B \in L(R^n)$ be symmetric positive definite. Show that the norm $\|\mathbf{x}\| = (\mathbf{x}^T B\mathbf{x})^{1/2}$ is monotonic if and only if $B = \alpha I$, $\alpha \neq 0$.

E3.2.10 Let

$$A = \begin{bmatrix} 2 & -10^{10} \\ 0 & 2 \end{bmatrix}$$

and $\mathbf{x}^T = (1, 10^{-10})^T$. Show that $\|A\mathbf{x} - \mathbf{x}\|_\infty = 10^{-10}$ and thus "conclude" that 1 is an "approximate" eigenvalue.

E3.2.11 Let λ and $\mathbf{x}$ be an approximate eigenpair of $A \in L(C^n)$ with $\|\mathbf{x}\|_2 = 1$. Set $\mathbf{r} = A\mathbf{x} - \lambda\mathbf{x}$. Show that λ and $\mathbf{x}$ are exact for some matrix $B \in L(C^n)$ such that $\mathrm{rank}(A - B) \leq 1$ and $\|A - B\|_2 \leq \|\mathbf{r}\|_2$.

3.3 SPECIAL RESULTS FOR SYMMETRIC MATRICES

If $A \in L(R^n)$ is symmetric, we know by 1.1.3 that A is orthogonally similar to a diagonal matrix, that is, $P^{-1}AP = \mathrm{diag}(\lambda_1, \ldots, \lambda_n)$ where P is orthogonal. Since $\|P\|_2 = \|P^{-1}\|_2 = 1$ for an orthogonal matrix, an immediate corollary of 3.2.6 and 3.2.7 is the following.

3.3.1 Let $A \in L(R^n)$ be symmetric. Then for any eigenvalue μ of $A + E$, there is an eigenvalue λ of A such that

$$|\lambda - \mu| \leq \|E\|_2. \tag{1}$$

Moreover, if $\|A\mathbf{x} - \bar{\lambda}\mathbf{x}\|_2 \leq \varepsilon$ with $\|\mathbf{x}\|_2 = 1$, then there is an eigenvalue λ of A such that $|\lambda - \bar{\lambda}| \leq \varepsilon$.

This result is a first indication of the fact that the eigenvalues of a symmetric matrix are well conditioned. Indeed, if A is symmetric, then a left eigenvector is just the transpose of the corresponding right eigenvector, and hence the condition number of 3.2.8 is unity.

If the eigenvalues of A are distinct and $\|E\|_2$ is sufficiently small, it also follows from 3.2.6 that the eigenvalues of A and $A + E$ can be ordered so that

$$|\lambda_i - \mu_i| \leq \|E\|_2, \qquad i = 1, \ldots, n. \tag{2}$$

But if E is also symmetric, it is an interesting fact that (2) holds regardless of the distinctness of the eigenvalues of A or the size of $\|E\|_2$. This will be a consequence of the following important theorem.

3.3.2   (Courant–Fischer Min–Max Representation)   Let $A \in L(R^n)$ be symmetric with eigenvalues $\lambda_n \geq \cdots \geq \lambda_1$. Then

$$\lambda_k = \min_{V_k} \max\{\mathbf{x}^T A \mathbf{x} : \mathbf{x} \in V_k, \quad \|\mathbf{x}\|_2 = 1\} \tag{3}$$

where V_k is an arbitrary subspace of R^n of dimension k.

Proof: Let $\mathbf{u}_1, \ldots, \mathbf{u}_n$ be an orthogonal set of eigenvectors of A with $\|\mathbf{u}_i\|_2 = 1$, $i = 1, \ldots, n$, and let M_k be the subspace spanned by $\mathbf{u}_k, \ldots, \mathbf{u}_n$. Then $\dim M_k = n - k + 1$ so that if V_k is any subspace of dimension k we must have $V_k \cap M_k \neq \{\mathbf{0}\}$. Therefore there is an $\mathbf{x} \in V_k \cap M_k$ with $\|\mathbf{x}\|_2 = 1$. Thus, with $\mathbf{x} = \sum_{i=k}^n \beta_i \mathbf{u}_i$,

$$\mathbf{x}^T A \mathbf{x} = \sum_{i=k}^n \lambda_i \beta_i^2 \geq \lambda_k \sum_{i=k}^n \beta_i^2 = \lambda_k$$

so that, since V_k was arbitrary,

$$\min \max \mathbf{x}^T A \mathbf{x} \geq \lambda_k.$$

To prove the reverse inequality, let V_k^0 be the subspace spanned by $\mathbf{u}_1, \ldots, \mathbf{u}_k$ and let $\mathbf{x} \in V_k^0$ with $\|\mathbf{x}\|_2 = 1$ and $\mathbf{x} = \sum_{i=1}^k \gamma_i \mathbf{u}_i$. Then

$$\mathbf{x}^T A \mathbf{x} = \sum_{i=1}^k \lambda_i \gamma_i^2 \leq \lambda_k,$$

so that

$$\max\{\mathbf{x}^T A \mathbf{x} : \mathbf{x} \in V_k^0, \quad \|\mathbf{x}\|_2 = 1\} \leq \lambda_k.$$

Hence, (3) follows. $\$\$\$$

As an easy consequence of the previous result, we obtain the afore-mentioned perturbation theorem.

3.3.3   (Symmetric Perturbation Theorem)   Let A and $A + E$ in $L(R^n)$ be symmetric with eigenvalues $\lambda_n \geq \cdots \geq \lambda_1$ and $\mu_n \geq \cdots \geq \mu_1$, respectively. Then

$$|\lambda_i - \mu_i| \leq \|E\|_2, \qquad i = 1, \ldots, n.$$

Proof: If $\gamma = \|E\|_2$, then $E + \gamma I$ is positive semidefinite. Hence, since $A + E + \gamma I$ has eigenvalues $\mu_i + \gamma$, 3.3.2 yields

$$\mu_j + \gamma = \min_{V_j} \max\{\mathbf{x}^T(A + E + \gamma I)\mathbf{x} : \mathbf{x} \in V_j, \quad \|\mathbf{x}\|_2 = 1\}$$

$$\geq \min_{V_j} \max\{\mathbf{x}^T A \mathbf{x} : \mathbf{x} \in V_j, \quad \|\mathbf{x}\|_2 = 1\} = \lambda_j$$

so that $\mu_j - \lambda_j \geq -\gamma$. Similarly, since $E - \gamma I$ is negative semidefinite,

$$\mu_j - \gamma = \min \max \mathbf{x}^T(A + E - \gamma I)\mathbf{x} \leq \min \max \mathbf{x}^T A \mathbf{x} = \lambda_j$$

so that $\mu_j - \lambda_j \leq \gamma$. $\$\$\$

Theorem 3.3.3 shows that small changes in the elements of the matrix A cause only correspondingly small changes in the eigenvalues of A. *Hence the eigenvalues of a real symmetric matrix are well conditioned.* It is important to realize, however, that this statement pertains to absolute changes in the eigenvalues and not relative changes. Consider, for example, changing the diagonal elements of $A = \mathrm{diag}(1, 10^{-1}, \ldots, 10^{-10})$ by 10^{-10}. Then the relative error in the eigenvalue 1 is 10^{-10}, but that in 10^{-10} is 1. This is reflected in the observation that a good computational procedure for the eigenvalues of a symmetric matrix should be able to obtain the largest eigenvalues with a low relative error, but that there are intrinsic difficulties in obtaining the lowest eigenvalues with a low relative error.

We next state without proof two famous theorems which complement and supplement 3.3.3.

3.3.4   (Hoffman–Wielandt Theorem)   Let A, $A + E$, and E in $L(R^n)$ be symmetric with eigenvalues $\lambda_n \geq \cdots \geq \lambda_1$, $\mu_n, \geq \cdots \geq \mu_1$, and $\gamma_n \geq \cdots \geq \gamma_1$, respectively, and define the vectors

$$\boldsymbol{\lambda} = (\lambda_1, \ldots, \lambda_n)^T, \qquad \boldsymbol{\mu} = (\mu_1, \ldots, \mu_n)^T, \qquad \boldsymbol{\gamma} = (\gamma_1, \ldots, \gamma_n)^T.$$

Then $\|\boldsymbol{\lambda} - \boldsymbol{\mu}\|_2 \leq \|\boldsymbol{\gamma}\|_2$.

For the next result, recall that the **convex hull** of a set of vectors is the smallest convex set which contains them.

3.3.5   (Lidskii–Wielandt Theorem)   Under the conditions of 3.3.4, λ lies in the convex hull of the set of vectors of the form $\mu + P\gamma$, where P runs over all possible permutation matrices.

We give an example of the use of this last theorem as contrasted with 3.3.3. Suppose that $n = 2$ and $\mu_1 = 1$, $\mu_2 = 2$, $\gamma_1 = -\varepsilon$, $\gamma_2 = 2\varepsilon$. Then $\|E\|_2 = |\gamma_2| = 2\varepsilon$ and by 3.3.3

$$|\lambda_1 - 1| \le 2\varepsilon, \qquad |\lambda_2 - 2| \le 2\varepsilon.$$

That is, the vector λ lies in a square with center $\mu = (1, 2)$ and side 4ε (see Figure 3.3.1). On the other hand, it is easy to see that the convex hull

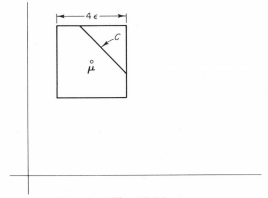

Figure 3.3.1

C of 3.3.5 is simply the line segment between the points $(1 - \varepsilon, 2 + 2\varepsilon)$ and $(1 + 2\varepsilon, 2 - \varepsilon)$ so that 3.3.5 gives a much more precise location of the eigenvalues of A. It is important to note, however, that both 3.3.4 and 3.3.5 require estimates for all eigenvalues of E, and not just $\|E\|_2$. They are therefore not too useful in practice.

We next obtain some results on the sensitivity of eigenvectors. We first give an a posteriori error bound.

3.3.6   (A Posteriori Estimate)   Let $A \in L(R^n)$ be symmetric with eigenvalues $\lambda_1, \ldots, \lambda_n$ and let $\|A\mathbf{x} - \lambda\mathbf{x}\|_2 \le \varepsilon$ for $\|\mathbf{x}\|_2 = 1$. Suppose that

$$|\lambda - \lambda_i| \ge d > 0, \qquad i \ne j. \tag{4}$$

Then A has a normalized eigenvector $\mathbf{u}_j$ corresponding to λ_j such that

$$\|\mathbf{x} - \mathbf{u}_j\|_2 \leq \gamma(1 + \gamma^2)^{1/2}, \tag{5}$$

where $\gamma = \varepsilon/d$.

Proof: Let $\mathbf{u}_1, \ldots, \mathbf{u}_n$ be normalized orthogonal eigenvectors corresponding to $\lambda_1, \ldots, \lambda_n$ and let

$$\mathbf{x} = \sum_{i=1}^{n} \alpha_i \mathbf{u}_i.$$

Without loss of generality we may assume that $\alpha_j \geq 0$ since otherwise we could work with $-\mathbf{u}_j$ instead of $\mathbf{u}_j$. Now

$$\varepsilon^2 \geq \|A\mathbf{x} - \lambda\mathbf{x}\|_2^2 = \sum_{i=1}^{n} (\lambda_i - \lambda)^2 \alpha_i^2 \geq d^2 \sum_{i \neq j} \alpha_i^2,$$

so that, since $\|\mathbf{x}\|_2 = 1$,

$$\alpha_j^2 = 1 - \sum_{i \neq j} \alpha_i^2 \geq 1 - \gamma^2. \tag{6}$$

Hence, since $\alpha_j \geq 0$,

$$(1 - \alpha_j)^2 = (1 - \alpha_j^2)^2/(1 + \alpha_j)^2 \leq (1 - \alpha_j^2)^2 \leq \gamma^4.$$

Thus

$$\|\mathbf{x} - \mathbf{u}_j\|_2^2 = (1 - \alpha_j)^2 + \sum_{i \neq j} \alpha_i^2 \leq \gamma^4 + \gamma^2$$

which is the desired result. $\$\$\$$

Note that for this result to be useful, ε should be small compared to d.

It is interesting that as a simple corollary of 3.3.6 we may obtain a result on the perturbation of eigenvectors. Recall (E3.1.5) that even if A is symmetric, the eigenvectors are not necessarily continuous functions of the elements of A unless the corresponding eigenvalues are simple. Hence, the following estimate involves the eigenvalue separation as well as $\|E\|$.

3.3.7 (Eigenvector Perturbation Theorem)   Let A and $A + E$ in $L(R^n)$ be symmetric with eigenvalues $\lambda_n \geq \cdots \geq \lambda_1$ and $\mu_n \geq \cdots \geq \mu_1$. If

$$|\lambda_i - \lambda_j| \geq \beta > \|E\|_2, \qquad i \neq j, \tag{7}$$

then A and $A + E$ have normalized eigenvectors $\mathbf{u}_j$ and $\mathbf{v}_j$ corresponding to λ_j and μ_j such that

$$\|\mathbf{u}_j - \mathbf{v}_j\|_2 \leq \gamma(1 + \gamma^2)^{1/2},$$

where $\gamma = \|E\|_2/(\beta - \|E\|_2)$.

Proof: By 3.3.3, together with (7), we have

$$|\lambda_i - \mu_j| \geq |\lambda_i - \lambda_j| - |\lambda_j - \mu_j| \geq \beta - \|E\|_2, \qquad i \neq j$$

while, clearly,

$$A\mathbf{v}_j - \mu_j \mathbf{v}_j = -E\mathbf{v}_j.$$

Hence 3.3.6 applies with $\mathbf{x} = \mathbf{v}_j$ and $\lambda = \mu_j$. $\$\$\$$

We consider next a useful procedure for improving an approximate eigenvalue. Let $A \in L(R^n)$ again be symmetric and assume that $\|\mathbf{x}\|_2 = 1$. Then the quantity

$$\lambda_R = \mathbf{x}^T A \mathbf{x} \tag{8}$$

is called a **Rayleigh quotient** for A. If $\mathbf{x}$ is a reasonable approximation to an eigenvector, then λ_R will be an excellent approximation to an eigenvalue. This is made precise in the following result.

3.3.8 (Rayleigh Quotient Theorem) Let $A \in L(R^n)$ be symmetric with eigenvalues $\lambda_1, \ldots, \lambda_n$, and for given $\mathbf{x}$ with $\|\mathbf{x}\|_2 = 1$ define λ_R by (8). Assume that $|\lambda_R - \lambda_i| \geq d > 0$, $i \neq j$, and set $\gamma = \varepsilon/d$, where $\varepsilon = \|A\mathbf{x} - \lambda_R \mathbf{x}\|_2$. Then

$$|\lambda_R - \lambda_j| \leq \varepsilon\gamma/(1 - \gamma^2).$$

Proof: Let

$$\mathbf{x} = \sum_{i=1}^{n} \alpha_i \mathbf{u}_i$$

where $\mathbf{u}_1, \ldots, \mathbf{u}_n$ are the normalized eigenvectors of A. Then

$$\lambda_R = \sum_{i=1}^{n} \lambda_i \alpha_i^2 = \lambda_R \sum_{i=1}^{n} \alpha_i^2,$$

the latter equality being trivial since $\sum \alpha_i^2 = 1$. Hence

$$|\lambda_j - \lambda_R|\alpha_j^2 \leq \sum_{i \neq j} |\lambda_i - \lambda_R|\alpha_i^2 \leq (1/d)\sum_{i \neq j} |\lambda_i - \lambda_R|^2 \alpha_i^2$$

$$\leq (1/d)\sum_{i=1}^{n} (\lambda_i - \lambda_R)^2 \alpha_i^2 = (1/d)\|A\mathbf{x} - \lambda_R \mathbf{x}\|_2^2 \leq \varepsilon\gamma.$$

Also, by (6),

$$\alpha_j^2 \geq 1 - \gamma^2,$$

so that

$$|\lambda_j - \lambda_R| \leq \frac{\varepsilon\gamma}{\alpha_j^2} \leq \frac{\varepsilon\gamma}{(1-\gamma^2)}. \quad \$\$\$$$

We consider an example of the previous theorems. Suppose that we have computed an approximate eigenvalue $\bar{\lambda}$ and eigenvector $\mathbf{x}$ of the symmetric matrix $A \in L(R^n)$, and that

$$\|A\mathbf{x} - \bar{\lambda}\mathbf{x}\|_2 \leq 10^{-8}, \qquad \|\mathbf{x}\|_2 = 1.$$

Then, by 3.3.1, $\bar{\lambda}$ is within 10^{-8} of a true eigenvalue of A. To apply 3.3.6 and 3.3.8 we need to obtain estimates on the separation of the eigenvalues, and we will assume, as is invariably the case, that this information is not known a priori. Suppose, instead, that we have computed all approximate eigenvalues and vectors $\bar{\lambda}_1 \geq \cdots \geq \bar{\lambda}_n$, and $\mathbf{x}_1, \ldots, \mathbf{x}_n$, and that

$$\min_{i \geq 2} |\bar{\lambda}_i - \bar{\lambda}_{i-1}| \geq 10^{-3} \tag{9}$$

while

$$\max_i \|A\mathbf{x}_i - \bar{\lambda}_i \mathbf{x}_i\|_2 \leq 10^{-8} \tag{10}$$

where we assume that $\|\mathbf{x}_i\|_2 = 1$. Hence, again by 3.3.1, each of the intervals

$$I_i = [\bar{\lambda}_i - 10^{-8}, \bar{\lambda}_i + 10^{-8}]$$

contains at least one true eigenvalue of A. More importantly, we can conclude by (9) that each I_i contains precisely one true eigenvalue λ_i; for (9) shows that the I_i are disjoint, so that if any I_i contained two true eigenvalues, then some I_j would have no eigenvalue which would contradict 3.3.1. Therefore, we may take the quantity d of (4) to be $d \geq 10^{-3} - 10^{-8}$. Theorem 3.3.6 now shows that for each computed eigenvector there is a normalized true eigenvector $\mathbf{u}_i$ which satisfies

$$\|\mathbf{x}_i - \mathbf{u}_i\|_2 \leq \frac{10^{-8}}{10^{-3} - 10^{-8}} \left[1 + \left(\frac{10^{-8}}{10^{-3} - 10^{-8}}\right)^2\right]^{1/2} \doteq 10^{-5}.$$

Finally, if we compute the Rayleigh quotients

$$\lambda_i^R = \mathbf{x}_i^T A \mathbf{x}_i, \qquad i = 1, \ldots, n$$

then E3.3.4 shows that (10) holds with $\tilde{\lambda}_i$ replaced by λ_i^R. Now assume that (9) holds for the λ_i^R. Then we may again take $d \geq 10^{-3} - 10^{-8}$ in 3.3.8 and obtain

$$|\lambda_i^R - \lambda_i| \leq \frac{(10^{-8})^2}{(10^{-3} - 10^{-8})(1 - 10^{-10})} \doteq 10^{-13}$$

a considerable improvement over the original estimates.

EXERCISES

E3.3.1   Prove Theorems **3.3.1**, **3.3.2**, and **3.3.3** for hermitian matrices in $L(C^n)$.

E3.3.2   Let H_n be the $n \times n$ Hilbert matrix of (2.1.3) and let $\bar{H}_n$ be the matrix obtained by truncating the elements of H_n to 27 binary digits, as discussed in Section 2.1. Give bounds for the deviation of the eigenvalues of $\bar{H}_n$ from those of H_n.

E3.3.3   Let $A \in L(R^n)$ be symmetric with eigenvalues $\lambda_1 \leq \cdots \leq \lambda_n$ and let A_{n-1} be the **leading principal submatrix** of A of order $n - 1$, obtained by deleting the last row and column of A. If $\mu_1 \leq \cdots \leq \mu_{n-1}$ are the eigenvalues of A_{n-1} show that $\lambda_1 \leq \mu_1 \leq \lambda_2 \leq \mu_2 \leq \cdots \leq \mu_{n-1} \leq \lambda_n$. (*Hint*: Use **3.3.2**.)

E3.3.4   Let $A \in L(R^n)$ be symmetric and $\mathbf{x}$ a fixed vector with $\|\mathbf{x}\|_2 = 1$. Show that $\lambda_R = \mathbf{x}^T A \mathbf{x}$ minimizes the function

$$f(\lambda) = \|A\mathbf{x} - \lambda\mathbf{x}\|_2.$$

E3.3.5   Let $A = \text{diag}(1, \ldots, n) + \varepsilon \mathbf{u}\mathbf{u}^T$ where $\mathbf{u}^T = (1, \ldots, n)$. Give the best estimates and error bounds that you can for the eigenvalues and eigenvectors of A.

READING

The basic references for this chapter are Householder [1964], Wilkinson [1963], and Wilkinson [1965].

CHAPTER 4

DIFFERENTIAL AND DIFFERENCE EQUATIONS

4.1 DIFFERENTIAL EQUATIONS

Consider the system of differential equations

$$y_1' = y_2, \qquad y_2' = y_1 \tag{1}$$

with the initial conditions

$$y_1(0) = 1, \qquad y_2(0) = -1 \tag{2}$$

and the exact solution

$$y_1(t) = e^{-t}, \qquad y_2(t) = -e^{-t}. \tag{3}$$

Now consider a slight change in the initial condition for y_1, say

$$\hat{y}_1(0) = 1 + \varepsilon, \qquad \hat{y}_2(0) = -1.$$

With this initial condition, the exact solution of (1) is

$$\hat{y}_1(t) = \tfrac{1}{2}(2 + \varepsilon)e^{-t} + \tfrac{1}{2}\varepsilon e^t, \qquad \hat{y}_2(t) = -\tfrac{1}{2}(2 + \varepsilon)e^{-t} + \tfrac{1}{2}\varepsilon e^t.$$

That is, an arbitrarily small change ε in the initial condition produces a change in the solution which grows exponentially; in particular,

$$|y_i(t) - \hat{y}_i(t)| \to +\infty \qquad \text{as} \quad t \to +\infty, \qquad i = 1, 2. \tag{4}$$

The solution of the initial value problem (1), (2) is unstable with respect to small changes in the initial conditions and the attempt to compute a numerical solution will be thwarted by this "ill conditioning."

There is a well-developed mathematical theory of stability of solutions of initial value problems. In this section, we shall give only some basic results.

4.1.1 Definition Let $f: R^n \times R^1 \to R^n$. Then a solution $\mathbf{y}$ of the differential equation

$$\mathbf{y}' = f(\mathbf{y}, t) \tag{5}$$

is **stable** (with respect to changes in initial conditions)† if given $\varepsilon > 0$ there is a $\delta > 0$ so that any other solution $\hat{\mathbf{y}}$ of (5) for which

$$\|\mathbf{y}(0) - \hat{\mathbf{y}}(0)\| \leq \delta \tag{6}$$

satisfies

$$\|\mathbf{y}(t) - \hat{\mathbf{y}}(t)\| \leq \varepsilon, \quad \text{if} \quad t \in (0, \infty). \tag{7}$$

Moreover, $\mathbf{y}$ is **asymptotically stable** if, in addition to being stable,

$$\|\mathbf{y}(t) - \hat{\mathbf{y}}(t)\| \to 0 \quad \text{as} \quad t \to +\infty.$$

Finally, $\mathbf{y}$ is **relatively stable** if (7) is replaced by

$$\|\mathbf{y}(t) - \hat{\mathbf{y}}(t)\| \leq \varepsilon \|\mathbf{y}(t)\|, \quad \text{if} \quad t \in (0, \infty). \quad \text{\$\$\$}$$

As we shall see, the last concept of relative stability is probably of the most importance to numerical analysis.

For linear equations with constant coefficients, stability can be characterized completely in terms of the spectrum of the coefficient matrix as the following basic theorem shows.

4.1.2 (Stability Theorem) A solution $\mathbf{y}$ of

$$\mathbf{y}' = A\mathbf{y} \tag{8}$$

is stable if and only if all eigenvalues of $A \in L(R^n)$ have nonpositive real part and any eigenvalue with zero real part belongs to a 1×1 Jordan block. Moreover, $\mathbf{y}$ is asymptotically stable if and only if all eigenvalues of A have negative real part.‡

† Also called "stable in the sense of Lyapunov."

‡ Note that for the linear system (8), a given solution $\mathbf{y}$ is stable if and only if every solution of (8) is stable.

Proof: We recall that the solution of (8) with initial condition $y(0) = y_0$ can be written in the form (see E4.1.2 and E4.1.3)

$$y(t) = e^{At}y_0.$$

Here, e^B is the matrix defined by the (always convergent) series

$$e^B = I + B + \frac{1}{2}B^2 + \frac{1}{3!}B^3 + \cdots. \tag{9}$$

If $\hat{y}$ is any other solution of (8), then $w = y - \hat{y}$ satisfies (8), and it suffices to show that the trivial solution of (8) is stable, i.e., that given $\varepsilon > 0$ there is a $\delta > 0$ so that whenever $\|w(0)\| \le \delta$, then $\|w(t)\| \le \varepsilon$ for all t. The solution w can be represented as

$$w(t) = e^{At}w(0)$$

and since $w(0)$ is arbitrary (except as to norm), it suffices to examine the matrix function e^{At}. Let $A = PJP^{-1}$ where J is the Jordan form of A. Then

$$e^{At} = e^{PJP^{-1}t} = Pe^{Jt}P^{-1},$$

and, using (9) and E1.1.11, if J_0 is any Jordan block of J, then

$$e^{J_0 t} = e^{\lambda t}\begin{bmatrix} 1 & t & \cdots & t^{m-1}/(m-1)! \\ & & \ddots & \vdots \\ & & & t \\ \text{\Large O} & & \ddots & 1 \end{bmatrix}$$

where m is the dimension of the block J_0. Both results are now evident. Clearly $e^{J_0 t} \to 0$ as $t \to +\infty$ if and only if Re $\lambda < 0$ while if Re $\lambda = 0$, then $e^{J_0 t}$ remains bounded if and only if $m = 1$. $\$\$\$

As an illustration of this theorem, we return to the equation (1) which may be written in the form (8) with

$$A = \begin{bmatrix} 0 & 1 \\ 1 & 0 \end{bmatrix}.$$

The eigenvalues of A are ± 1 and hence no solution of (1) is stable.

As another example, we consider the nth-order equation

$$y^{(n)} + a_{n-1}y^{(n-1)} + \cdots + a_0 y = 0 \tag{10}$$

for which initial conditions will be prescribed for $y(0), y'(0), \ldots, y^{(n-1)}(0)$. In the usual way, we change (10) to a system by introducing the variables $y_1 = y, y_2 = y', \ldots, y_n = y^{(n-1)}$. Then (10) is equivalent to (8) with

$$A = \begin{bmatrix} 0 & 1 & 0 & \cdots & & 0 \\ & 0 & 1 & \ddots & & \vdots \\ & & \ddots & \ddots & & 0 \\ & & & 0 & & 1 \\ -a_0 & -a_1 & & \cdots & & -a_{n-1} \end{bmatrix}. \tag{11}$$

[Note that the system (1) arises from the equation $y'' = y$ in this way.] We will need the following lemma about the matrix A of (11).

4.1.3† Let A be of the form (11) and let λ be an eigenvalue of multiplicity m. Then there is only one Jordan block associated with λ, and it is of dimension m.

Proof: The first $n - 1$ rows of A are linearly independent, and the same is true of $A - \mu I$ for any value of μ; hence rank$(A - \lambda I) = n - 1$. Since the Jordan form J and A are similar, they must have the same rank and therefore rank$(J - \lambda I) = n - 1$. But if more than one Jordan block were associated with λ we would have rank$(J - \lambda I) < n - 1$ which would be a contradiction. $\$\$\$

According to Definition 4.1.1, the solution of (10) is stable if and only if not only y but also all its derivatives up to order $n - 1$ are bounded under small perturbations. It may seem more satisfying to adopt the definition that y is stable if and only if $|y^{(i)}(0) - \hat{y}^{(i)}(0)| \leq \delta$, $i = 0, \ldots,$ $n - 1$, implies that $|y(t) - \hat{y}(t)| \leq \varepsilon$ for all $t \geq 0$. However, it is easy to see, for linear equations, that this is equivalent to 4.1.1 (see E4.1.4). Hence, we have the following corollary of Theorem 4.1.2 together with 4.1.3.

4.1.4 Any solution of (10) is stable if and only if the eigenvalues of the matrix (11) have nonpositive real part and any eigenvalue with zero real part is simple.

From the point of view of numerical analysis the type of stability in the above results is not as important as relative stability. Consider again the system (1) with the general solution

$$y_1(t) = c_1 e^{-t} + c_2 e^t, \qquad y_2(t) = -c_1 e^{-t} + c_2 e^t.$$

† Alternative formulation: Let A be of the form (11) and let $\lambda_1, \ldots, \lambda_p$ be the distinct eigenvalues of A. Then the Jordan form of A has precisely p blocks.

These solutions are easy to compute numerically by the usual methods such as Runge–Kutta or predictor–corrector provided that $c_2 \neq 0$ (or very small). It is true that for large t, the absolute error in the computed solution will be large but the relative error can be kept much smaller. However, if $c_2 = 0$, then the computed solution will be contaminated by unwanted contributions from e^t, and it is extremely difficult to compute a solution with small relative error. The solution with $c_2 = 0$ is relatively unstable while any solution with $c_2 \neq 0$ is relatively stable. This is an immediate consequence of the following general theorem for linear equations with constant coefficients.

4.1.5   (Relative Stability Theorem) For $A \in L(R^n)$, let Λ denote the set of eigenvalues of A with maximal real part. Then a solution $\mathbf{y}$ of the linear system $\mathbf{y}' = A\mathbf{y}$, $\mathbf{y}(0) = \mathbf{y}^0$, is relatively stable if and only if there is some $\lambda \in \Lambda$ such that $\mathbf{y}(0)$ has a component in the direction of a principal vector of maximal degree associated with λ.

Proof: Again write $\mathbf{y}(t) = e^{At}\mathbf{y}^0$ and $\hat{\mathbf{y}}(t) = e^{At}\hat{\mathbf{y}}^0$, and consider first the case in which A has diagonal canonical form. Let $\mathbf{u}_1, \ldots, \mathbf{u}_n$ be linearly independent eigenvectors corresponding to the eigenvalues $\lambda_1, \ldots, \lambda_n$. Then e^{At} has eigenvalues $e^{\lambda_i t}$ and eigenvectors $\mathbf{u}_i$ (E4.1.5) so that

$$\frac{\|\mathbf{y}(t) - \hat{\mathbf{y}}(t)\|}{\|\mathbf{y}(t)\|} = \frac{\|e^{At}(\mathbf{y}^0 - \hat{\mathbf{y}}^0)\|}{\|e^{At}\mathbf{y}^0\|} = \frac{\|\sum \alpha_i e^{\lambda_i t}\mathbf{u}_i\|}{\|\sum \beta_i e^{\lambda_i t}\mathbf{u}_i\|}. \tag{12}$$

The dominant terms in numerator and denominator are $\alpha_j e^{\lambda_j t}\mathbf{u}_j$ and $\beta_j e^{\lambda_j t}\mathbf{u}_j$, where $\lambda_j \in \Lambda$. Now suppose that $\beta_j = 0$ for all j such that $\lambda_j \in \Lambda$. Then, since $\hat{\mathbf{y}}^0$ is arbitrary in direction, the corresponding α_j may be chosen to be nonzero and, clearly, the quantity of (12) tends to ∞. On the other hand, if any $\beta_j \neq 0$, $\lambda_j \in \Lambda$, then the quantity is bounded and indeed may be made arbitrarily small with α_i. That is, it is necessary and sufficient that $\mathbf{y}^0$ have a component in the direction of some eigenvector associated with an eigenvalue in Λ.

Next consider the general case. The dominant term in the denominator will now have terms containing $e^{\lambda_j t}t^{m-1}/(m-1)!$, where m is the dimension of the largest Jordan block of any eigenvalue in Λ, and it is necessary and sufficient that the coefficients of these dominant terms do not all vanish. This is precisely the condition that $\mathbf{y}(0)$ has a component in the direction of a principal vector of maximal degree. $\$\$\$$

The previous theorems have all been for linear equations. For non-linear equations, the possible behavior is much more varied, and there may even be instability because solutions have a singularity for finite t. Consider, for example, the equation

$$y' = ty(y - 2), \qquad y(0) = y_0.$$

The exact solution is

$$y(t) = \frac{2y_0}{y_0 + (2 - y_0)e^{t^2}}.$$

Hence, for $y_0 = 2$, $y(t) \equiv 2$, while for $y_0 < 2$, $y(t)$ goes asymptotically to zero as $t \to +\infty$. However, for $y_0 > 2$, $y(t)$ has a singularity when $y_0 + 2(-y_0)e^{t^2} = 0$ (see Figure 4.1.1).

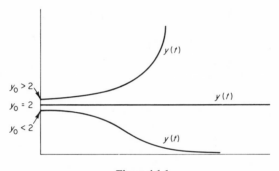

Figure 4.1.1

It is possible, however, to give a local version of Theorem 4.1.2.

4.1.6 Consider the equation

$$\mathbf{y}' = A\mathbf{y} + f(\mathbf{y}) \tag{13}$$

where all the eigenvalues of $A \in L(R^n)$ have negative real parts. Assume that $f: R^n \to R^n$ is continuous in a neighborhood of $\mathbf{0}$ and that

$$\lim_{\mathbf{y}\to 0} \|f(\mathbf{y})\|/\|\mathbf{y}\| = 0.$$

Then there is a $\delta > 0$ such that for any $\mathbf{y}(0)$ with $\|\mathbf{y}(0)\| \le \delta$, the solution of (13) satisfies

$$\lim_{t \to \infty} \mathbf{y}(t) = \mathbf{0}.$$

We will not prove this theorem; its proof may be found in several places (see, e.g., Coddington and Levinson [1955, p. 314]). Its analogue for difference equations (Theorem 4.2.7) will be of particular importance to us.

EXERCISES

E4.1.1 Consider the differential equations

(a) $y'' + 3y' + 2y = 0$
(b) $y'' + y' = 0$
(c) $4y'' - y = 0$

with initial conditions $y(0) = y_0$, $y'(0) = y_1$. Give the stable, asymptotically stable, and relatively stable solutions as functions of y_0 and y_1.

E4.1.2 Prove that the series (9) converges for any $B \in L(C^n)$.

E4.1.3 Verify both formally and rigorously that $\mathbf{y}(t) = e^{At}\mathbf{y}_0$ is the unique solution of $\mathbf{y}' = A\mathbf{y}$, $\mathbf{y}(0) = \mathbf{y}_0$.

E4.1.4 Show that a solution of (10) is stable if and only if for given $\varepsilon > 0$ there is a $\delta > 0$ so that $|\hat{y}^{(i)}(0) - y^{(i)}(0)| \leq \delta, i = 0, \ldots, n - 1$ implies that $|\hat{y}(t) - y(t)| \leq \varepsilon$ for all $t \geq 0$.

E4.1.5 Let $A \in L(R^n)$ have eigenvalues $\lambda_1, \ldots, \lambda_n$ and eigenvectors $\mathbf{x}_1, \ldots, \mathbf{x}_n$. Show that e^A has eigenvalues e^{λ_i} and eigenvectors $\mathbf{x}_i, i = 1, \ldots, n$.

4.2 DIFFERENCE EQUATIONS

We consider now results for difference equations analogous to the results of the previous section for differential equations. These results will, by and large, be of much more interest to us and, as will be seen in Part III, actually provide convergence theorems for iterative processes.

In analogy with the differential equation (4.1.5), consider the first-order vector difference equation

$$\mathbf{y}^{k+1} = G(\mathbf{y}^k, k), \qquad k = 0, 1, \ldots \tag{1}$$

where $G: R^n \times R^1 \to R^n$ is a given mapping and $\mathbf{y}^0$ is assumed known. We now paraphrase the definitions and results of the previous section in terms of (1).

4.2.1 Definition Let $G: R^n \times R^1 \to R^n$. Then a solution $\{\mathbf{y}^k\}$ of the difference equation (1) is

(a) **stable** if given $\varepsilon > 0$ there is a $\delta > 0$ so that if $\{\hat{\mathbf{y}}^k\}$ is another solution of (1) for which $\|\mathbf{y}^0 - \hat{\mathbf{y}}^0\| \le \delta$ then

$$\|\mathbf{y}^k - \hat{\mathbf{y}}^k\| \le \varepsilon, \qquad k = 1, \ldots \tag{2}$$

(b) **asymptotically stable** if, in addition,

$$\|\mathbf{y}^k - \hat{\mathbf{y}}^k\| \to 0 \qquad \text{as} \quad k \to \infty \tag{3}$$

(c) **relatively stable** if (2) is replaced by

$$\|\mathbf{y}^k - \hat{\mathbf{y}}^k\| \le \varepsilon \|\mathbf{y}^k\|, \qquad k = 1, \ldots . \tag{4}$$

Note that this definition assumes that the solutions of (1) actually exist.

We next give an analogue of Theorem 4.1.2 for the linear difference equation with constant coefficients

$$\mathbf{y}^{k+1} = B\mathbf{y}^k + \mathbf{d}, \qquad k = 0, 1, \ldots, \qquad \mathbf{y}^0 \text{ given.} \tag{5}$$

4.2.2 (Stability Theorem) Let $B \in L(R^n)$ and $\mathbf{d} \in R^n$. Then a solution $\{\mathbf{y}^k\}$ of (5) is

(a) stable if and only if $\rho(B) \le 1$ and if $\rho(B) = 1$, then B is of class† M

(b) asymptotically stable if and only if $\rho(B) < 1$.

Proof: (a) Let $\{\hat{\mathbf{y}}^k\}$ be any other solution of (5) and set $\mathbf{w}^k = \hat{\mathbf{y}}^k - \mathbf{y}^k$, $k = 0, 1, \ldots$. Then

$$\mathbf{w}^k = B\mathbf{w}^{k-1} = \cdots = B^k\mathbf{w}^0, \qquad k = 0, 1, \ldots . \tag{6}$$

† See Definition 1.3.7.

Suppose that $\{B^k\}$ is bounded, that is, that $\|B^k\| \le \sigma$, $k = 0, 1, \dots$. Then by (6), $\|\hat{\mathbf{y}} - \mathbf{y}^k\| \le \varepsilon$ for all $\hat{\mathbf{y}}^0$ such that

$$\|\hat{\mathbf{y}}^0 - \mathbf{y}^0\| \le \delta = \varepsilon/\sigma.$$

Conversely, if $\{B^k\}$ is not bounded, then $\|B^k\mathbf{x}\| \to \infty$ as $k \to \infty$ for some $\mathbf{x} \in R^n$. Hence if $\hat{\mathbf{y}}^0$ is chosen so that $\mathbf{y}^0 - \hat{\mathbf{y}}^0$ is in the direction $\mathbf{x}$, it follows that $\|\mathbf{y}^k - \hat{\mathbf{y}}^k\| \to +\infty$ as $k \to \infty$ so that the solution is not stable. Therefore, the result follows from Theorem 1.3.9.

(b) Proceeding as in (a), it is easy to see that the solution is asymptotically stable if and only if $B^k \to 0$ as $k \to \infty$ and, by Theorem 1.3.9, this is true if and only if $\rho(B) < 1$. $\$\$\$$

We note that for the linear equation (2), asymptotic stability is equivalent to **global** asymptotic stability; that is, (3) holds for any $\hat{\mathbf{y}}^0$. We also note that a result similar to 4.1.5 for relative stability is easily proved (E4.2.2). Finally, we note that 4.2.2 contains the basic convergence theorem for (5), when (5) is considered as an iterative process. We shall return to this topic in Part III.

Theorem 4.2.2 corresponds exactly to Theorem 4.1.2 with the difference that the condition on the real part of the eigenvalues in 4.1.2 becomes in 4.2.2 a condition on the spectral radius. This is quite natural; for consider the differential equation

$$\mathbf{y}' = A\mathbf{y} \tag{7}$$

and Euler's method for its numerical integration:

$$\mathbf{y}^{k+1} = \mathbf{y}^k + hA\mathbf{y}^k \equiv B\mathbf{y}^k, \qquad B = I + hA. \tag{8}$$

We expect intuitively that, for sufficiently small h, the stability properties of (7) and (8) will be the same. The following simple result shows that this is indeed true for asymptotic stability. Note, however, that it is not necessarily true for stability (E4.2.3).

4.2.3 Let $A \in L(R^n)$. Then the eigenvalues of A all have negative real part if and only if there is a $h_0 > 0$ such that for all $0 < h \le h_0$, $\rho(I + hA) < 1$.

Proof: If $\rho(I + hA) < 1$, then all eigenvalues of hA lie in a disk centered at -1 and with radius <1; that is, all eigenvalues of hA have negative real part and since $h > 0$ the same is true of A. Conversely, suppose that all eigenvalues of A have negative real part. Since there are only finitely many, they lie in some square with vertices $-a \pm ib$ and $-(a + b) \pm ib$ and hence the eigenvalues of $I + hA$ lie in a square with vertices $1 + h(-a \pm ib)$ and $1 + h(-a - b \pm ib)$. Thus for $\rho(I + hA) < 1$, it suffices that

$$(1 - ah)^2 + h^2 b^2 < 1 \qquad \text{and} \qquad [1 - (a + b)h]^2 + h^2 b^2 < 1$$

and these inequalities can both be satisfied for sufficiently small h. $\$\$\$$

We consider next an nth-order linear difference equation of the form

$$y_k - \alpha_{n-1} y_{k-1} - \cdots - \alpha_0 y_{k-n} = 0, \qquad k = n, n+1, \ldots \qquad (9)$$

with given initial conditions $y_0, \ldots, y_{n-1}$. We convert this to a vector difference equation by defining

$$\mathbf{y}^k = \begin{bmatrix} y_{k+n-1} \\ \vdots \\ y_k \end{bmatrix}, \qquad B = \begin{bmatrix} \alpha_{n-1} & \cdots & & \alpha_0 \\ 1 & 0 & \cdots & 0 \\ & & \ddots & \ddots & \vdots \\ \mathbf{O} & & & 1 & 0 \end{bmatrix}. \qquad (10)$$

Then, clearly, (9) is equivalent to the difference equation

$$\mathbf{y}^{k+1} = B\mathbf{y}^k, \qquad k = 0, 1, \ldots. \qquad (11)$$

Hence, as a corollary of **4.2.2**, we have the following result (see also **E4.2.7**).

4.2.4 Let $B \in L(R^n)$ be given by (10). Then the solution of (11) is stable if and only if all roots λ_i of the polynomial

$$p(\lambda) = \lambda^n - \alpha_{n-1} \lambda^{n-1} - \cdots - \alpha_0 \qquad (12)$$

satisfy $|\lambda_i| \le 1$ and if $|\lambda_i| = 1$, then λ_i is a simple root. The solution is asymptotically stable if and only if all $|\lambda_i| < 1$.

Proof: Since B is the companion matrix of $p(\lambda)$, **4.1.3** shows that if λ_i is an eigenvalue of multiplicity $m > 1$ then there is a Jordan block of dimension m associated with λ, so that B is not of class M. Hence, the result is a direct consequence of **4.2.2**. $\$\$\$$

Suppose that the roots $\lambda_1, \ldots, \lambda_n$ of (12) are distinct. Then the general solution of (9) may be written in the form

$$y_k = \sum_{i=1}^{n} c_i \lambda_i^k, \qquad k = 0, 1, \ldots \qquad (13)$$

for arbitrary constants $c_1, \ldots, c_n$. This is an immediate consequence of the fact that the solution of (11) is

$$\mathbf{y}^k = B^k \mathbf{y}^0 = P^{-1} J^k P \mathbf{y}^0, \qquad k = 0, 1, \ldots \tag{14}$$

where J is the Jordan form of B. Since the eigenvalues are assumed to be distinct, J is diagonal and hence $J^k P y^0$ is of the form $\mathbf{d}^k = (\lambda_1^k b_1, \ldots, \lambda_n^k b_n)^{\mathrm{T}}$. By (10), y_k is then the inner product of the last row of P^{-1} and $\mathbf{d}^k$, which shows that the solution is of the form (13).

The above analysis holds for any initial conditions $y_0, \ldots, y_{n-1}$. However, if these are prescribed, then $c_1, \ldots, c_n$ are completely determined. Indeed, (13) for $k = 0, \ldots, n - 1$, gives the system of equations

$$
\begin{aligned}
y_0 &= c_1 + \cdots + c_n \\
y_1 &= \lambda_1 c_1 + \cdots + \lambda_n c_n \\
&\ \ \vdots \\
y_{n-1} &= \lambda_1^{n-1} c_1 + \cdots + \lambda_n^{n-1} c_n
\end{aligned}
$$

for the unknown constants in terms of $y_0, \ldots, y_{n-1}$ and $\lambda_1, \ldots, \lambda_n$. The coefficient matrix is the famous Vandermonde matrix which is nonsingular since the λ_i are distinct.

More generally, if the polynomial (12) has distinct roots $\lambda_1, \ldots, \lambda_m$ with multiplicities $n_1, \ldots, n_m$, then 4.1.3 shows that the Jordan form J of B has precisely m blocks J_i that are either (λ_i) or of the form

$$
J_i = \begin{bmatrix}
\lambda_i & 1 & & \\
& \cdot & \cdot & \\
& & \cdot & 1 \\
& & & \lambda_i
\end{bmatrix}.
$$

Hence, following the analysis above,

$$
y_k = (\mathbf{a}_1, \ldots, \mathbf{a}_m) \begin{bmatrix}
J_1^k \mathbf{b}_1 \\
\vdots \\
J_m^k \mathbf{b}_m
\end{bmatrix} \tag{15}
$$

where $\mathbf{a}$ is the last row of P^{-1} and $\mathbf{b} = P\mathbf{y}^0$, both partitioned according to the structure of J, and, by E1.1.11,

$$
J_i^k \mathbf{b}_i = \begin{bmatrix}
\lambda_i^k & k\lambda_i^{k-1} & \cdots & \binom{k}{n_i-1}\lambda_i^{k-n_i+1} \\
& \cdot \cdot & & \vdots \\
\mathbf{O} & & & \lambda_i^k
\end{bmatrix}
\begin{bmatrix}
b_{i1} \\
\vdots \\
b_{in_i}
\end{bmatrix}
= \begin{bmatrix}
\sum_{j=0}^{n_i-1} \binom{k}{j} b_{ij} \lambda_i^{k-j} \\
\vdots \\
\lambda_i^k b_{in_i}
\end{bmatrix}
$$

Therefore, (15) shows that the solution y_k is a linear combination of

$$\lambda_1^k, k\lambda_1^{k-1}, \ldots, \binom{k}{n_1-1}\lambda_1^{k-n_1+1}, \lambda_2^k, \ldots, \lambda_m^k, \ldots, \binom{k}{n_m-1}\lambda_m^{k-n_m+1}$$

which, since $\binom{k}{i}$ is just a polynomial in k, is equivalent to taking a linear combination of

$$\lambda_1^k, k\lambda_1^k, \ldots, k^{n_1-1}\lambda_1^k, \lambda_2^k, \ldots, \lambda_m^k, \ldots, k^{n_m-1}\lambda_m^k.$$

The important thing to note is that various powers of k now appear, due to the multiplicity of the roots.

We turn next to the linear difference equation

$$\mathbf{y}^{k+1} = B\mathbf{y}^k + \mathbf{d}^k, \qquad k = 0, 1, \ldots \tag{16}$$

with a nonconstant sequence $\{\mathbf{d}^k\}$ of vectors. If $\mathbf{d}^k \equiv \mathbf{0}$, the solution of (16) may be written as $\mathbf{y}^k = B^k\mathbf{y}^0$. More generally, we can obtain an expression for the solution of (16) by the following calculation:

$$\mathbf{y}^k = B(B\mathbf{y}^{k-2} + \mathbf{d}^{k-2}) + \mathbf{d}^{k-1} = B^2\mathbf{y}^{k-2} + B\mathbf{d}^{k-2} + \mathbf{d}^{k-1}$$

$$= \cdots = B^k\mathbf{y}^0 + \sum_{j=0}^{k-1} B^j\mathbf{d}^{k-j-1}. \tag{17}$$

This exact formula is frequently useful. However, norm estimates which derive from it are often more useful. One such result is the following.

4.2.5 Let $B \in L(R^n)$. If the zero solution of the homogeneous equation

$$\mathbf{y}^{k+1} = B\mathbf{y}^k, \qquad k = 0, 1, \ldots \tag{18}$$

is stable, then, in some norm, the solution of (16) satisfies

$$\|\mathbf{y}^k\| \le \|\mathbf{y}^0\| + \sum_{j=0}^{k-1} \|\mathbf{d}^j\|, \qquad k = 1, \ldots. \tag{19}$$

If the zero solution of (18) is asymptotically stable, then there is a norm and a constant $\alpha < 1$ so that the solution of (16) satisfies

$$\|\mathbf{y}^k\| \le \alpha^k\|\mathbf{y}^0\| + \sum_{j=0}^{k-1} \alpha^j \|\mathbf{d}^j\|, \qquad k = 1, \ldots. \tag{20}$$

Proof: If the zero solution of (18) is stable, then **4.2.2** shows that either $\rho(B) < 1$ or that $\rho(B) = 1$ and B is of class M. In either case we have by **1.3.6** or **1.3.8** that there is a norm such that $\|B\| \le 1$ and (19) follows directly from (17). In case the zero solution of (18) is asymptotically stable, then **4.2.2** shows that $\rho(B) < 1$ and hence, by **1.3.6**, $\|B\| < 1$ in some norm. Therefore, with $\alpha = \|B\|$, (20) again follows directly from (17).   \$\$\$

A corollary of **4.2.5** which will be useful in Chapter 5 is the following.

4.2.6   Assume that the roots $\lambda_1, \ldots, \lambda_n$ of the polynomial (12) satisfy $|\lambda_i| \leq 1$, $i = 1, \ldots, n$, and any λ_i for which $|\lambda_i| = 1$ is simple. Then there is a constant $c \geq 1$ such that any solution of

$$y_k - \alpha_{n-1} y_{k-1} - \cdots - \alpha_0 y_{k-n} = \gamma_k, \qquad k = n, n+1, \ldots \qquad (21)$$

where $\{\gamma_k\}$ is a given sequence, satisfies

$$|y_k| \leq c\left(\max_{0 \leq i \leq n-1} |y_i| + \sum_{j=n}^{k} |\gamma_j| \right), \qquad k = n, n+1, \ldots. \qquad (22)$$

Proof: By the representation (10), we can write (21) in the form (16) where $\mathbf{d}^k = (\gamma_{k+n-1}, 0, \ldots, 0)^{\mathrm{T}}$, and, by **4.2.5**, there is a norm such that

$$\|\mathbf{y}^{k-n+1}\| \leq \|\mathbf{y}^0\| + \sum_{j=0}^{k-n} \|\mathbf{d}^j\|, \qquad k = n, n+1, \ldots \qquad (23)$$

But by the norm-equivalence theorem **1.2.4** there are constants $c_2 \geq c_1 > 0$ so that

$$c_1 \|\mathbf{y}\|_\infty \leq \|\mathbf{y}\| \leq c_2 \|\mathbf{y}\|_\infty .$$

Therefore, (22) follows from (23) with $c = c_2/c_1$.   \$\$\$

We end this section by considering the nonlinear difference equation

$$\mathbf{y}^{k+1} = B\mathbf{y}^k + f(\mathbf{y}^k) \qquad (24)$$

where f is "small" for small $\mathbf{y}$ in the sense that

$$\lim_{y \to 0} \frac{\|f(\mathbf{y})\|}{\|\mathbf{y}\|} = 0. \qquad (25)$$

4.2.7   (Perron's Theorem)   Assume that $B \in L(R^n)$ with $\rho(B) < 1$ and that $f: R^n \to R^n$ is continuous in a neighborhood of $\mathbf{0}$ and satisfies (25). Then the zero solution of (24) is asymptotically stable.

Proof: By **1.3.6**, we can choose a norm for which $\|\mathbf{B}\| = \alpha < 1$. Let $\gamma > 0$ be such that $\alpha + \gamma < 1$. Then, in this norm, there is a $\delta > 0$ so that

$$\|f(\mathbf{y})\| \leq \gamma \|\mathbf{y}\| \qquad \text{if} \quad \|\mathbf{y}\| \leq \delta.$$

Hence, for any $\mathbf{y}^0$ such that $\|\mathbf{y}^0\| \leq \delta$, we have

$$\|\mathbf{y}^1\| \leq \|B\| \, \|\mathbf{y}^0\| + \|f(\mathbf{y}^0)\| \leq (\alpha + \gamma) \|\mathbf{y}^0\| \leq \delta.$$

It follows by induction that

$$\|\mathbf{y}^k\| \leq (\alpha + \gamma)^k \|\mathbf{y}^0\| \leq \delta, \qquad k = 1, 2, \ldots$$

so that $\mathbf{y}^k \to 0$ as $k \to \infty$. $\$\$\$$

EXERCISES

E4.2.1 Consider the difference equations

(a) $y_k + 3y_{k-1} + 2y_{k-2} = 0$

(b) $4y_k - y_{k-2} = 0$.

Give the stable, asymptotically stable, and relatively stable solutions as functions of the initial conditions y_0 and y_1. Contrast your results with E4.1.1.

E4.2.2 Show that a solution $\{\mathbf{y}^k\}$ of (5) is relatively stable if and only if $\mathbf{y}^0$ has a component in the direction of a principal vector of maximal degree associated with an eigenvalue with modulus equal to $\rho(B)$.

E4.2.3 Show that any solution of the system

$$\mathbf{y}' = A\mathbf{y}, \qquad A = \begin{bmatrix} 0 & 1 \\ -1 & 0 \end{bmatrix}$$

is stable but that no solution of the corresponding difference equation (8) for Euler's method is stable.

E4.2.4 Let $B \in L(R^4)$ be the matrix of E3.2.2. For any $\mathbf{d} \in R^4$, show that every solution of (5) is globally asymptotically stable.

E4.2.5 Let $B \in L(R^n)$ satisfy $\rho(B) < 1$ and consider the difference equation (16) where $\mathbf{y}^0$ is given and $\{\mathbf{d}^k\}$ is a given sequence satisfying $\|\mathbf{d}^k\| \leq \delta$, $k = 0, 1, \ldots$. Show that there are constants c_1, c_2, α with $\alpha < 1$, such that

$$\|\mathbf{y}^k\|_2 \leq c_1 \alpha^k + c_2, \qquad k = 0, 1, \ldots.$$

E4.2.6 Verify directly that (13) is a solution of (9).

E4.2.7 Define a solution of (9) to be stable if given $\varepsilon > 0$ there is a $\delta > 0$ so that $|\hat{y}_i - y_i| \leq \delta, i = 1, \ldots, n - 1$ implies that $|y_i - \hat{y}_i| \leq \varepsilon, i = n, n + 1, \ldots$. Show that this is equivalent to 4.2.1 applied to (10) and (11).

READING

Stability theorems for ordinary differential equations are discussed in numerous books; see for example, Coddington and Levinson [1955] and Hahn [1967]. Corresponding results for difference equations are not as readily available in textbook literature; see, however, Hahn [1967].

The concept of relative stability for either differential or difference equations does not seem to have been previously explored.

PART II

DISCRETIZATION ERROR

We now begin our study of the three basic kinds of error in numerical analysis. The first, to be considered in this part, is discretization error. This error results, in general, whenever we must approximate a function defined on a continuum of points by one which is specified by only finitely many parameters. For example, in Chapter 5, we consider initial value problems for ordinary differential equations. Here, a typical procedure is to evaluate the function f of the equation $y' = f(y)$ at a finite number of grid points in the interval of integration and combine these values in one way or the other to obtain an approximate solution at the grid points. The fact that the solution depends on the values of f and y at all points of the interval implies that, except in certain trivial cases, the approximate solution will be in error. The task is then to obtain estimates for this discretization error and to show that the error tends to zero as the distance between the grid points tends to zero. The problem is similar for the solution of boundary value problems to be treated in Chapter 6; an added difficulty here is that evaluation of the approximate solution requires, in general, the solution of a nontrivial system of equations.

CHAPTER 5

DISCRETIZATION ERROR
FOR INITIAL VALUE PROBLEMS

5.1 CONSISTENCY AND STABILITY

In this chapter, we will treat the initial value problem

$$y'(x) = f(x, y(x)), \qquad a \le x \le b, \quad y(a) = y_0 \tag{1}$$

and the general class of integration methods

$$y_{k+n} = \alpha_{n-1}y_{k+n-1} + \cdots + \alpha_0 y_k + h\Phi(x_{k+n}, \ldots, x_k; y_{k+n}, \ldots, y_k; h). \tag{2}$$

Here $x_i = a + ih$, $i = 1, \ldots, N$, where $h = (b - a)/N$ is the constant **steplength**; $\alpha_0, \ldots, \alpha_{n-1}$ are given constants; Φ is a given function which depends on f, although we have not indicated this dependence explicitly; and the initial values $y_0, \ldots, y_{n-1}$ are assumed to be known. For simplicity, we shall assume that (1) is a scalar equation although many of our results hold for systems of ordinary differential equations.

If $n = 1$, then (2) is called a **one-step method**; otherwise, it is a **multistep method** or, more precisely, an **n-step method**. If Φ is independent of y_{k+n}, then the method is **explicit**; in this case, y_{k+n} may be computed simply by evaluating the right-hand side of (2), while if the method is **implicit** a (generally nonlinear) equation must be solved to obtain y_{n+k}.

While (2) is not the most general procedure possible, it does contain as special cases most of the usual methods of interest. We shall review some of these briefly. The simplest method is that of **Euler**:

$$y_{k+1} = y_k + hf_k, \qquad k = 0, 1, \ldots;$$

here, and henceforth, f_k will denote $f(x_k, y_k)$. More complicated one-step methods are those of **Heun** (also called the **second-order Runge–Kutta method**):

$$y_{k+1} = y_k + \tfrac{1}{2}h[f_k + f(x_{k+1}, y_k + hf_k)] \qquad (3)$$

and the (fourth-order) **Runge–Kutta method**:

$$y_{k+1} = y_k + \tfrac{1}{6}[\Gamma_1 + 2\Gamma_2 + 2\Gamma_3 + \Gamma_4] \qquad (4)$$

where

$$\Gamma_1 = hf_k, \qquad\qquad\qquad \Gamma_2 = hf(x_k + \tfrac{1}{2}h, y_k + \tfrac{1}{2}\Gamma_1)$$

$$\Gamma_3 = hf(x_k + \tfrac{1}{2}h, y_k + \tfrac{1}{2}\Gamma_2), \qquad \Gamma_4 = hf(x_k + h, y_k + \Gamma_3).$$

In the case of Euler's method, the function Φ of (2) is simply f itself, while for (3) it is

$$\Phi(x; y; h) = \tfrac{1}{2}[f(x, y) + f(x + h; y + hf(x, y))]. \qquad (5)$$

It is left to E5.1.1 to give Φ for the method (4).

The above methods are all explicit. A simple example of an implicit one-step method is the **trapezoidal rule**:

$$y_{k+1} = y_k + \tfrac{1}{2}h[f_k + f(x_{k+1}, y_{k+1})]. \qquad (6)$$

Note that both this method and (3) reduce to the usual trapezoidal rule for numerical integration when f is independent of y.

A very important special case of (2) is the **general linear multistep method**

$$y_{k+n} = \alpha_{n-1}y_{k+n-1} + \cdots + \alpha_0 y_k + h[\beta_n f_{n+k} + \cdots + \beta_0 f_k] \qquad (7)$$

where $\alpha_0, \ldots, \alpha_{n-1}$ and $\beta_0, \ldots, \beta_n$ are given constants. (The method is called linear because the f_i appear linearly—in contrast to the Runge–Kutta methods, for example—*not* because the differential equation is linear.) If $\beta_n = 0$ the method is explicit and otherwise is implicit. As well-known examples of (7), we note the (fourth-order) **Adams–Bashforth method**

$$y_{k+4} = y_{k+3} + (h/24)[55f_{k+3} - 59f_{k+2} + 37f_{k+1} - 9f_k] \qquad (8)$$

which is explicit, and the implicit (fourth-order) **Adams–Moulton method**

$$y_{k+3} = y_{k+2} + (h/24)[9f_{k+3} + 19f_{k+2} - 5f_{k+1} + f_k]. \qquad (9)$$

Recall that (8) and (9) are usually used in conjunction as a "predictor–corrector" pair. Another pair of formulas of interest to us is that of **Milne**:

$$y_{k+3} = y_{k-1} + (4h/3)[2f_{k+2} - f_{k+1} + 2f_k] \qquad (10)$$

and

$$y_{k+2} = y_k + (h/3)[f_{k+2} + 4f_{k+1} + f_k]. \qquad (11)$$

Note that in all of these multistep methods, the supplementary initial conditions $y_1, \ldots, y_{n-1}$ must be supplied by some auxiliary mechanism, usually a one-step method such as (4).

Let y be the solution of the differential equation (1) on the interval $[a, b]$ and $y_0, \ldots, y_N$, with $Nh = b - a$, the approximate solution given by (2). Then the quantity

$$\max_{0 \le i \le N} |y(x_i) - y_i| \qquad (12)$$

is the **global discretization error** of the approximate solution. Our main concern in this chapter will be to obtain an estimate of this discretization error and to show that the approximate solution converges to the true solution as the steplength h tends to zero. As a first step toward this goal, we introduce a quantity which is easier to compute than (12).

5.1.1 Definition Assume that (1) has a solution y on the interval $[a, b]$. Then the **local discretization error**† of the method (2) at $x \in [a, b - nh]$ is given by

$$\tau(x, h) = h^{-1}[y(x + nh) - \alpha_{n-1}y(x + (n-1)h) - \cdots - \alpha_0 y(x)]$$
$$- \Phi(x + nh, x + (n-1)h, \ldots, x; y(x + nh), \ldots, y(x); h). \qquad (13)$$

Note that $\tau(x, h)$ is simply the "residual," divided by h, when the exact solution y is substituted into the expression (2). The first term of $\tau(x, h)$ is, in some sense, an approximation to $y'(x)$ while the second is an approximation to f. The sense in which Φ and the constants $\alpha_0, \ldots, \alpha_{n-1}$ define suitable approximations is made precise by the following.

5.1.2 Definition Assume that (1) has a solution y on the interval $[a, b]$. Then the method (2) is **consistent** with the differential equation (1) if $\lim_{h \to 0} \tau(h) = 0$ where

$$\tau(h) = \max_{x \in [a, b - nh]} |\tau(x, h)|. \qquad (14)$$

† Some authors define the local discretization error to be $h\tau(x, h)$. See also E5.1.2.

The following simple result gives algebraic conditions on Φ and $\alpha_0, \ldots,$ α_{n-1} in order that the method be consistent.

5.1.3   Assume that (1) has a continuously differentiable solution y on the interval $[a, b]$ and that the function Φ of (2) is continuous in all of its arguments. Define the polynomial

$$p(\lambda) = \lambda^n - \alpha_{n-1}\lambda^{n-1} - \cdots - \alpha_0 \tag{15}$$

and assume that

$$p(1) = 0 \tag{16}$$

and that

$$\Phi(x, \ldots, x; y(x), \ldots, y(x); 0) = p'(1)f(x, y(x)) \tag{17}$$

for all $x \in [a, b]$. Then the method (2) is consistent with the equation (1). Moreover, if $y \not\equiv 0$, then (16) and (17) are also necessary for consistency.

Proof: If we set

$$g(x, h) = \frac{y(x + h) - y(x)}{h} - y'(x)$$

and

$$G(x, h) = ng(x, nh) - \sum_{j=1}^{n-1} j\alpha_j g(x, jh)$$

then we may write $\tau(x, h)$ as

$$\tau(x, h) = \frac{1}{h}p(1)y(x) + p'(1)y'(x) + G(x, h)$$
$$- \Phi(x + nh, \ldots, x; y(x + nh), \ldots, y(x); h). \tag{18}$$

The first term is zero by (16), and $G(x, h) \to 0$ as $h \to 0$ by the differentiability of y. Therefore, since $y'(x) = f(x, y(x))$, (17) and the continuity of y and Φ show that

$$\lim_{h \to 0} \tau(x, h) = y'(x)p'(1) - \Phi(x, \ldots, x; y(x), \ldots, y(x); 0) = 0.$$

Moreover, it is easy to see that the continuity of y' on the compact interval $[a, b]$ ensures that the convergence of $G(x, h)$ to zero is uniform for $x \in [a, b]$ and, similarly, the continuity of Φ and y as a function of their arguments ensures that the convergence of Φ to $y'(x)p'(1)$ is uniform for

$x \in [a, b)$. Hence $\tau(x, h) \to 0$ as $h \to 0$ uniformly for $x \in [a, b)$ and thus $\tau(h) \to 0$ as $h \to 0$.

Conversely, if the method is consistent, then (13) shows that $p(1)y(x) = 0$ for all $x \in [a, b]$. Thus if $y \not\equiv 0$, then (16) holds and, since $G(x, h) \to 0$ as $h \to 0$, (18) ensures that (17) is valid. $\$\$\$$

Since $p(\lambda) = \lambda - \alpha_0$ for a one-step method, 5.1.3 shows that a consistent one-step method must have the form

$$y_{k+1} = y_k + h\Phi(x_{k+1}, x_k; y_{k+1}, y_k; h).$$

Moreover, since $p'(1) = 1$, (17) implies that also

$$\Phi(x, x; y(x), y(x); 0) = f(x, y(x)).$$

It is easy to verify that all of the particular methods (3)–(6) satisfy this condition. For the linear multistep method (7), the condition (17) reduces to

$$\sum_{i=0}^{n} \beta_i = n - (n - 1)\alpha_{n-1} - \cdots - \alpha_1$$

and, again, it is easy to show that the methods (8)–(11) are consistent.

It is tempting to suppose that the consistency of a method will ensure that the global discretization error tends to zero with h. Rather surprisingly, this indeed turns out to be true for one-step methods but not necessarily for multistep methods.

Consider the method†

$$y_{k+2} = 4y_{k+1} - 3y_k - 2hf_k,$$

which is readily seen, by 5.1.3, to be consistent. If we apply this method to the differential equation $y' = y$ with $y(0) = 1$, then it becomes the linear difference equation

$$y_{k+2} = 4y_{k+1} - (3 + 2h)y_k, \qquad y_0 = 1, \qquad y_1 = \text{known.} \qquad (19)$$

This difference equation is easily solved (E5.1.4) in the form (4.2.13), and we obtain

$$y_k = y_k(h) = -\frac{(y_1 - 2 - \sqrt{1 - 2h})}{2\sqrt{1 - 2h}} (2 - \sqrt{1 - 2h})^k$$

$$+ \frac{(y_1 - 2 + \sqrt{1 - 2h})}{2\sqrt{1 - 2h}} (2 + \sqrt{1 - 2h})^k. \qquad (20)$$

† This method arises in a natural way by differentiation of an interpolating polynomial; for the derivation, see Henrici [1962, p. 219].

Now take $y_1 = e^h$, and let k tend to infinity and h tend to zero in such a way that $x = kh$ remains fixed. Then it is easy to see (E5.1.4) that the first term of (20) converges to e^x, the exact solution of the differential equation, but that the second term tends to infinity; that is, the solution of (19) does not converge to the solution of the differential equation as h tends to zero, for the particular choice of y_1. On the other hand, if we take $y_1 = 2 - \sqrt{1 - 2h}$, then the second term of (20) is zero and the first term still tends to e^x as $h \to 0$. However, this is the only value of y_1 for which the second term is zero and for any other y_1 this term will ultimately dominate. In other words, the solution

$$y_k = (2 - \sqrt{1 - 2h})^k$$

of (19), which approximates the solution of the differential equation, is relatively unstable in the sense of 4.2.1.

The above discussion shows that an additional stability condition is necessary in order that a consistent method be convergent as $h \to 0$. It turns out that the condition is just stability of the linear part of (2).

5.1.4   Definition   The method (2) is **stable** if every solution of the difference equation

$$y_{k+n} - \alpha_{n-1} y_{k+n-1} - \cdots - \alpha_0 y_k = 0$$

is stable.†

By 4.2.4, we can restate this stability condition as follows.

5.1.5   (Root Condition)   The method (2) is stable if and only if the roots $\lambda_1, \ldots, \lambda_n$ of the polynomial (15) satisfy $|\lambda_i| \leq 1$, $i = 1, \ldots, n$ and any root of modulus 1 is simple.

Note that the method (19) is not stable since the roots of $\lambda^2 - 4\lambda + 3 = 0$ are 1 and 3. Note also that any consistent one-step method is stable.

In the next section, we will show that for a stable and consistent method, the global discretization error tends to zero with h. It is important to realize, however, that this alone does not necessarily imply a satisfactory numerical procedure. Consider, for example, the method

$$y_{k+2} = y_k + 2hf_{k+1}. \tag{21}$$

† Recall that the stability of solutions of this difference equation is defined by 4.2.1, in terms of the system (4.2.10), (4.2.11).

It is an immediate consequence of 5.1.3 and 5.1.5 that this method is both stable and consistent. Now apply (21) to the differential equation

$$y'(x) = -y(x), \qquad y(0) = 1 \tag{22}$$

so that (21) becomes

$$y_{k+2} = -2hy_{k+1} + y_k, \qquad y_0 = 1, \qquad y_1 = \text{known.} \tag{23}$$

The solution of this linear difference equation is of the form

$$y_k = c_1 \lambda_1^k + c_2 \lambda_2^k \tag{24}$$

with

$$\lambda_1 = -h + \sqrt{1 + h^2}, \qquad \lambda_2 = -h - \sqrt{1 + h^2}$$

and

$$c_1 = \frac{1}{2} + \frac{y_1 + h}{2\sqrt{1 + h^2}}, \qquad c_2 = \frac{1}{2} - \frac{(y_1 + h)}{2\sqrt{1 + h^2}}.$$

Now note that, for any $h > 0$, $|\lambda_2| > 1$ so that, except for the choice of y_1 which makes $c_2 = 0$, $|y_k| \to \infty$ as $k \to \infty$. Hence, since the exact solution of (22) is e^{-x}, the approximate solution given by (23) will deviate arbitrarily far from the exact solution for sufficiently large x.

Since the root λ_1 above tends to 1 as $h \to 0$, the above behavior would be prevented if the root λ_2 had a limit strictly less than one in modulus. This leads to the following class of methods.

5.1.6 Definition The method (2) is **strongly stable** if the roots of the polynomial (15) satisfy $|\lambda_i| \le 1$, $i = 1, \ldots, n$ with strict inequality for $n - 1$ roots.

A stable method which is not strongly stable is sometimes called **weakly stable** or **weakly unstable**.

Since 5.1.3 shows that for a consistent method p must have a root equal to unity, the roots of p for a consistent strongly stable method must satisfy

$$\lambda_1 = 1, \qquad |\lambda_i| < 1, \qquad i = 2, \ldots, n.$$

In particular, any consistent one-step method is strongly stable, as are the Adams methods (8) and (9). The Milne methods (10) and (11), however, are only weakly stable (E5.1.6), and may exhibit unstable behavior numerically, especially for problems with decreasing solutions.

EXERCISES

E5.1.1   Give the function Φ of (2) for the methods (4) and (6)–(11).

E5.1.2 Another definition of the local discretization error is the following. Let y_{k+n} be given by (2) and assume that $y_k, \ldots, y_{k+n-1}$ are exact, that is, $y_i = y(x_i)$, $i = k, \ldots, k+n-1$, where y is the solution of (1). Then define

$$E(x_{k+n}, h) = |y_{k+n} - y(x_{k+n})|.$$

Show that $E(x_{k+n}, h) = h|\tau(x_k, h)|$.

E5.1.3   Show that the methods (3), (4), (6), and (8)–(11) are consistent.

E5.1.4   Show that (20) is the exact solution of (19). Now hold $x = kh$ fixed and let $h \to 0$ and $k \to \infty$. Show that the first term of (20) converges to e^x but that the second term diverges to $-\infty$.

E5.1.5   Write a computer program to find the solution $\{y_k\}$ of (19) [directly from (19)]. Experiment with different values of h and y_1, and comment on the accuracy of your results as approximations to e^x. Do the same for (23).

E5.1.6   Show that the Milne methods (10) and (11) are weakly unstable. Write a computer program to verify numerically that they may show unstable behavior.

5.2 CONVERGENCE AND ORDER

In the previous section, we considered the initial value problem

$$y'(x) = f(x, y(x)), \qquad a \le x \le b, \quad y(a) = y_0 \tag{1}$$

and various stability and consistency criteria for the method

$$y_{k+n} = \alpha_{k-1} y_{k+n-1} + \cdots + \alpha_0 y_k + h\Phi(x_{k+n}, \ldots, x_k; y_{k+n}, \ldots, y_k; h). \tag{2}$$

We shall now prove the following basic theorem which shows that stability together with consistency are sufficient to imply that the global discretization error converges to zero with h.

5.2.1 (Consistency plus Stability Implies Convergence)   Assume that the method (2) is stable and that the function Φ satisfies

$$|\Phi(t_0, \ldots, t_n ; u_0, \ldots, u_n ; h) - \Phi(t_0, \ldots, t_n ; v_0, \ldots, v_n ; h)|$$
$$\leq c \max_{0 \leq i \leq n} |u_i - v_i| \tag{3}$$

for all $t_1, \ldots, t_n$ in the interval $[a, b]$, all $u_0, \ldots, u_n, v_0, \ldots, v_n$, and all $h \geq 0$, where c is independent of h. Assume that (1) has a solution y on $[a, b]$ and denote the solution of (2) as a function of h by $y_i(h)$, $i = 0, 1, \ldots$. Then there are constants c_1 and c_2, independent of h, such that

$$|y(a + kh) - y_k(h)| \leq c_1 r(h) + c_2 \tau(h), \qquad k = n, n + 1, \ldots, (b - a)/h \tag{4}$$

where τ is given by (5.1.14) and

$$r(h) = \max_{0 \leq k \leq n - 1} |y(a + kh) - y_k(h)| \tag{5}$$

with $y_0(h), \ldots, y_{n-1}(h)$ the given initial conditions of (2). In particular, if $r(h) \to 0$ as $h \to 0$ and the method is consistent, then for each fixed $x \in [a, b]$

$$\lim_{\substack{k \to \infty \\ h \to 0}} y_k(h) = y(x) \tag{6}$$

where the limit in (6) is taken so that $a + kh = x$ remains fixed.

Proof: For fixed h, set $e_i = y_i(h) - y(x_i)$. Then, adding $h\tau(x_k, h)$ to (2), where $\tau(x, h)$ is given by (5.1.13), we obtain

$$e_{k+n} = \alpha_{n-1} e_{k+n-1} + \cdots + \alpha_0 e_k + \gamma_{k+n}, \qquad k = 0, 1, \ldots \tag{7}$$

where

$$\gamma_{k+n} = h[\Phi(x_{k+n}, \ldots, x_k ; y_{k+n}, \ldots, y_k ; h)$$
$$- \Phi(x_{k+n}, \ldots, x_k ; y(x_{k+n}), \ldots, y(x_k); h)] - h\tau(x_k, h). \tag{8}$$

If we regard the γ_i in (7) as known, then **4.2.6** ensures that there is a constant $\beta \geq 1$ such that

$$|e_{k+n}| \leq \beta \left[\max_{0 \leq i \leq n-1} |e_i| + \sum_{i=0}^{k} |\gamma_{i+n}| \right], \qquad k = 0, 1, \ldots.$$

Since from (3) and (8) we have that

$$|\gamma_{i+n}| \le h\left[c \max_{0 \le j \le n} |e_{i+j}| + |\tau(x_i, h)|\right]$$

it follows that

$$|e_{k+n}| \le \beta r(h) + \beta(k + 1)h\left[c \max_{0 \le i \le k+n} |e_i| + \tau(h)\right]. \tag{9}$$

Now set

$$\omega_k = \max_{0 \le i \le k} |e_i|, \qquad k = 0, 1, \ldots.$$

Then (9) implies that

$$\omega_{k+n} \le \beta r(h) + \beta c(k + 1)h\omega_{k+n} + \beta(k + 1)h\tau(h). \tag{10}$$

For given h, we now assume that k is allowed to become only so large that

$$(k + 1)h \le (2\beta c)^{-1} \equiv \delta. \tag{11}$$

If this is true, then, by (10), we have

$$\omega_{k+n} \le [1 - \beta c(k + 1)h]^{-1}[\beta r(h) + \beta(k + 1)h\tau(h)]$$
$$\le 2\beta[r(h) + \delta\tau(h)]$$

so that

$$|e_{k+n}| \le 2\beta[r(h) + \delta\tau(h)], \qquad (k + 1)h \le \delta. \tag{12}$$

This gives an estimate for the interval $I_1 = [a, a + \delta]$, where δ is independent of h. Now consider the next interval $I_2 = [a + \delta, a + 2\delta]$. We may assume that the "initial points" lie in I_1 and hence the errors for these "initial values" can be bounded by (12). Therefore we obtain

$$|e_{k+n}| \le 2\beta\{2\beta[r(h) + \delta\tau(h)] + \delta\tau(h)\}$$

for all k such that the grid points lie in I_2. Since the interval $[a, b]$ may be covered by finitely many intervals of length δ, it is clear that this procedure may be repeated to obtain an estimate of the form (4). $\$\$\$$

In the previous section, we gave several examples of stable and consistent methods. Hence, to apply 5.2.1 to these methods it remains only to verify (3), which is invariably a consequence of a Lipschitz condition on the function f of (1). That is, assume that f satisfies

$$|f(x, u) - f(x, v)| \le L|u - v| \qquad \text{for all } x \in [a, b] \text{ and all } u, v \tag{13}$$

and consider the general linear multistep method (5.1.7) where Φ is given by

$$\Phi(t_0, \ldots, t_n; u_0, \ldots, u_n; h) \equiv \sum_{i=0}^{n} \beta_i f(t_i, u_i).$$

Then

$$|\Phi(t_0, \ldots, t_n; u_0, \ldots, u_n; h) - \Phi(t_0, \ldots, t_n; v_0, \ldots, v_n; h)|$$

$$\leq \sum_{i=0}^{n} |\beta_i| \, |f(t_i, u_i) - f(t_i, v_i)| \leq L \sum_{i=0}^{n} |\beta_i| \, |u_i - v_i|$$

$$\leq \left(L \sum_{i=0}^{n} |\beta_i| \right) \max_{0 \leq i \leq n} |u_i - v_i|.$$

As another example, consider the one-step method (5.1.3) for which Φ is given by

$$\Phi(t; u; h) = \tfrac{1}{2}[f(t, u) + f(t + h, u + hf(t, u))]. \tag{14}$$

Then, again under the assumption of (13),

$$|\Phi(t; u; h) - \Phi(t; v; h)|$$

$$\leq \tfrac{1}{2}[| f(t, u) - f(t, v)|$$

$$+ |f(t + h, u + hf(t, u)) - f(t + h, v + hf(t, v))|]$$

$$\leq \tfrac{1}{2}[L|u - v| + L|u + hf(t, u) - v - hf(t, v)|]$$

$$\leq L|u - v| + \tfrac{1}{2}Lh|u - v| \leq L(1 + \tfrac{1}{2}h_0)|u - v|$$

where h_0 is the largest value of h to be considered.

We note that 5.2.1 can be sharpened in regard to the condition (3) by requiring that this hold only in some suitable neighborhood of the solution of (1). A corresponding weakening of (13) may then be made; in particular, it suffices that the partial derivative f_y be continuous in a neighborhood of the solution curve of (1).

An important aspect of the discretization error is the rate at which it tends to zero as h tends to zero. A first measure of this rate is given by the following condition on the local discretization error $\tau(h)$. Recall that the notation $\tau(h) = O(h^p)$ means that $h^{-p}\tau(h)$ is bounded as $h \to 0$.

5.2.2 Definition The method (2) is at least **order p** if, for any differential equation of the form (1) with a p-times continuously differentiable solution, $\tau(h) = O(h^p)$. The method is exactly order p if, in addition, $\tau(h) \neq O(h^{p+1})$

for some differential equation (1) with a $(p+1)$-times continuously differential solution.

Although the order of the method is defined in terms of the local discretization error, it can be linked to the global discretization error by means of 5.2.1.

5.2.3 Assume that the conditions of 5.2.1 are satisfied for a differential equation (1) with a p-times continuously differentiable solution, that $r(h) = O(h^p)$, and that the method (2) is at least of order p. Then for h tending to zero so that kh remains fixed, the global discretization error satisfies

$$|y(a + kh) - y_k(h)| = O(h^p).$$

The proof is an immediate consequence of (4) together with 5.2.2.

We next compute the order of some of the methods of the previous section. Consider first the Heun method (5.1.3), for which the function Φ is given by (14), and assume that f is twice continuously differentiable as a function of both its variables so that the Taylor expansion

$$f(x + h, y + q) = f(x, y) + f_x(x, y)h + f_y(x, y)q + O(h^2 + q^2) \qquad (15)$$

holds. Here the $O(h^2 + q^2)$ term arises rigorously by bounding the second partial derivatives of f in a suitable region of interest. In the sequel such necessary boundings will be assumed tacitly.

The assumption on f implies that the solution y of (1) is three times continuously differentiable and hence (14) and (15) yield

$$\begin{aligned}
\tau(x, h) &= (1/h)[y(x + h) - y(x)] - \Phi(x; y(x); h) \\
&= (1/h)[y'(x)h + \tfrac{1}{2}y''(x)h^2 + O(h^3)] - \tfrac{1}{2}[f(x, y(x)) + f(x, y(x)) \\
&\quad + f_x(x, y(x))h + f_y(x, y(x))hf(x, y(x))] + O(h^2 + (hf(x, y(x))^2)) \\
&= O(h^2)
\end{aligned}$$

since $y'(x) = f(x, y(x))$ and

$$y''(x) = \frac{d}{dx} f(x, y(x)) = f_x(x, y(x)) + f_y(x, y(x))y'(x).$$

It follows that $\tau(h) = O(h^2)$, and hence the method is at least of order 2. (The fact that the method is exactly of order 2 is left to E5.2.2.) We note that a similar, but more cumbersome, calculation shows that the Runge–Kutta method (5.1.4) is of order 4.

We consider next the linear multistep method (5.1.7):

$$y_{k+n} = \alpha_{n-1} y_{k+n-1} + \cdots + \alpha_0 y_k + h[\beta_n f_{n+k} + \cdots + \beta_0 f_k]. \tag{16}$$

Assume that the solution y of (1) is four times continuously differentiable so that the Taylor expansions

$$y(x + jh) = y(x) + y'(x)jh + \tfrac{1}{2}y''(x)(jh)^2 + \tfrac{1}{6}y'''(x)(jh)^3 + O(h^4)$$

and

$$y'(x + jh) = y'(x) + y''(x)jh + \tfrac{1}{2}y'''(x)(jh)^2 + O(h^3)$$

hold. Then recalling again that $y'(x) = f(x, y(x))$, we have

$$\begin{aligned}
\tau(x, h) &= (1/h)[y(x + nh) - \alpha_{n-1}y(x + nh - h) - \cdots - \alpha_0 y(x)] \\
&\quad - [\beta_n y'(x + nh) + \cdots + \beta_0 y'(x)] \\
&= (1/h)\{y(x) + y'(x)nh + \tfrac{1}{2}y''(x)n^2h^2 + \tfrac{1}{6}y'''(x)n^3h^3 + O(h^4) \\
&\quad - \alpha_{n-1}[y(x) + y'(x)(n - 1)h + \tfrac{1}{2}y''(x)(n - 1)^2h^2 \\
&\quad + \tfrac{1}{6}y'''(x)(n - 1)^3h^3 + O(h^4)] \\
&\qquad \vdots \\
&\quad - \alpha_1[y(x) + y'(x)h + \tfrac{1}{2}y''(x)h^2 + \tfrac{1}{6}y'''(x)h^3 + O(h^4)] - \alpha_0 y(x)\} \\
&\quad - \beta_n[y'(x) + y''(x)nh + \tfrac{1}{2}y'''(x)n^2h^2 + O(h^3)] \\
&\qquad \vdots \\
&\quad - \beta_1[y'(x) + y''(x)h + \tfrac{1}{2}y'''(x)h^2 + O(h^3)] - \beta_0 y'(x) \\
&= (1/h)y(x)(1 - \alpha_{n-1} - \cdots - \alpha_0) \\
&\quad + y'(x)[n - \alpha_{n-1}(n - 1) - \cdots - \alpha_1 - \beta_n - \cdots - \beta_0] \\
&\quad + y''(x)h\{\tfrac{1}{2}[n^2 - \alpha_{n-1}(n - 1)^2 - \cdots - \alpha_1] - \beta_n n - \cdots - \beta_1\} \\
&\quad + y'''(x)h^2\{\tfrac{1}{6}[n^3 - \alpha_{n-1}(n - 1)^3 - \cdots - \alpha_1] - \tfrac{1}{2}[\beta_n n^2 - \cdots - \beta_1]\} \\
&\quad + O(h^3).
\end{aligned}$$

Now suppose that we wish the method to be at least of order 3. Then all of the above terms except $O(h^3)$ must vanish independently of the solution y. This leads to the conditions

$$\alpha_0 + \cdots + \alpha_{n-1} = 1, \qquad \beta_0 + \cdots + \beta_n = n - (n - 1)\alpha_{n-1} - \cdots - \alpha_1 \tag{17}$$

$$\beta_n n + \beta_{n-1}(n - 1) + \cdots + \beta_1 = \tfrac{1}{2}[n^2 - \alpha_{n-1}(n - 1)^2 - \cdots - \alpha_1] \tag{18}$$

$$\tfrac{1}{2}[\beta_n n^2 + \beta_{n-1}(n - 1)^2 + \cdots + \beta_1] = \tfrac{1}{6}[n^3 - \alpha_{n-1}(n - 1)^3 - \cdots - \alpha_1]. \tag{19}$$

Note that (17) is simply the consistency condition of 5.1.3.

It is clear that if the solution y is suitably differentiable, we may carry out the above expansions to higher order to obtain relations analogous to (17)–(19). We summarize this in the following result.

5.2.4 The linear multistep method (16) is at least of order p if (17) holds and in addition

$$\beta_n n^{j-1} + \beta_{n-1}(n-1)^{j-1} + \cdots + \beta_1 = (1/j)[n^j - \alpha_{n-1}(n-1)^j - \cdots - \alpha_1],$$

$$j = 2, \ldots, p. \quad (20)$$

The method is exactly of order p if it is at least of order p, but (20) does not hold for $j = p + 1$.

We give a simple example of the above result. Consider the method

$$y_{k+2} = y_k + 2hf_{k+1},$$

already discussed in the previous section in another context. Here $n = 2$, $\alpha_0 = 1$, $\alpha_1 = 0$, $\beta_2 = \beta_0 = 0$, and $\beta_1 = 2$. Hence

$$\alpha_0 + \alpha_1 = 1, \qquad \beta_2 + \beta_1 + \beta_0 = 2$$

and

$$2\beta_2 + \beta_1 = \tfrac{1}{2}(4 - \alpha_1)$$

so that the method is at least of order 2. However,

$$2^2\beta_2 + \beta_1 \neq \tfrac{1}{2}(2^3 - \alpha_1)$$

so that it is not of order 3; that is, the method is exactly order 2. We leave to E5.2.3 the verification that the Adams and Milne methods (5.1.8)–(5.1.11) are exactly order 4.

The conditions (17) and (20) may be viewed as a linear system of equations in the α_i and β_i. If $p = 2n$, then there are $2n + 1$ equations, and these equations may be solved to obtain the $2n + 1$ parameters $\alpha_0, \ldots, \alpha_{n-1}$ and $\beta_0, \ldots, \beta_n$.

5.2.5 The parameters $\alpha_0, \ldots, \alpha_{n-1}$ and $\beta_0, \ldots, \beta_n$ may be chosen so that the method (16) is at least of order $2n + 1$.

It turns out that this result is of little consequence, however, in view of the following remarkable result.

5.2.6 (Dahlquist's Theorem)† Any n-step method of the form (16) of order greater than $n + 2$ is unstable. More precisely, if n is even, then the highest-order stable method is of order $n + 2$; while if n is odd, the highest-order stable method is of order $n + 1$.

As an example of the use of **5.2.6**, consider the Adams–Moulton method (5.1.9):

$$y_{k+3} = y_{k+2} + (h/24)[9f_{k+3} + 19f_{k+2} - 5f_{k+1} + f_k].$$

It is easy (**E5.2.3**) to verify that this method is of order 4. But since $n = 3$ is odd, **5.2.6** shows that the maximum order of a three-step method is 4 so that, in the sense of order, this is already an optimal three-step method.

EXERCISES

E5.2.1 Apply Theorem **5.2.1** to the methods (5.1.3)–(5.1.11) under the assumption (13).

E5.2.2 Show that the method (5.1.3) is exactly order 2.

E5.2.3 Show that the methods (5.1.8)–(5.1.11) are all exactly order 4.

READING

The proof of the basic theorem **5.2.1** follows that given in Isaacson and Keller [1966]. A much more detailed treatment of the material of this chapter may be found in Henrici [1962]. For more recent results on stability and references to the literature, see Gear [1971].

---

† The proof of this as well as **5.2.5** may be found, for example, in Henrici [1962].

CHAPTER 6

DISCRETIZATION ERROR
FOR BOUNDARY VALUE PROBLEMS

6.1 THE MAXIMUM PRINCIPLE

In this chapter we consider boundary value problems for ordinary differential equations. As sketched in the introduction, the finite difference method for the boundary value problem

$$y''(x) = f(x, y(x), y'(x)), \quad a \le x \le b, \quad y(a) = \alpha, \quad y(b) = \beta \quad (1)$$

leads to the system of equations

$$y_{i-1} - 2y_i + y_{i+1} = h^2 f\left(x_i, y_i, \frac{y_{i+1} - y_{i-1}}{2h}\right), \quad i = 1, \ldots, n \quad (2)$$

where $x_i = a + ih$, $h = (b - a)/(n + 1)$, and $y_0 = \alpha$, $y_{n+1} = \beta$.

Provided that the boundary problem (1) has a unique solution, the first problem is to ensure that the system (2) has also. Assuming that this is true and denoting this solution by $y_i(h)$, $i = 0, \ldots, n + 1$, then we wish to show that the global discretization error tends to zero; that is,

$$y_i(h) \to y(x_i) \quad \text{as} \quad h \to 0 \quad (3)$$

where, as usual, $x_i = a + ih$ is fixed as h tends to zero and i tends to infinity. Finally, one also wishes to know the rate of convergence of (3) in terms of h.

We will give complete answers to the above questions only for the linear problem

$$y''(x) = p(x)y'(x) + q(x)y(x) + r(x), \quad y(a) = \alpha, \quad y(b) = \beta \quad (4)$$

where p, q, and r are given continuous functions and

$$q(x) \geq 0, \qquad x \in [a, b]. \tag{5}$$

We will also assume that

$$h|p(x)| < 2, \qquad x \in [a, b] \tag{6}$$

which is simply a condition that h is sufficiently small relative to p. Note that if $p \equiv 0$, then (6) imposes no restriction upon h.

For (4), the difference equations (2) become

$$y_{i-1} - 2y_i + y_{i+1} = h^2\left[\frac{p_i}{2h}(y_{i+1} - y_{i-1}) + q_i y_i + r_i\right], \qquad i = 1, \ldots, n \tag{7}$$

where we have set $p_i = p(x_i)$, and similarly for r_i and q_i. If we collect terms and set

$$a_i = (2 + q_i h^2), \qquad b_i = 1 + \tfrac{1}{2}p_i h, \qquad c_i = 1 - \tfrac{1}{2}p_i h \tag{8}$$

the equations (7) become

$$-b_i y_{i-1} + a_i y_i - c_i y_{i+1} = -r_i h^2, \qquad i = 1, \ldots, n \tag{9}$$

which, since y_0 and y_{n+1} are known, is a linear system of equations in the n unknowns $y_1, \ldots, y_n$.

Now note that (6) implies that

$$b_i > 0, \qquad c_i > 0, \qquad b_i + c_i \leq a_i, \qquad i = 1, \ldots, n. \tag{10}$$

These conditions suffice to ensure a unique solution for the system (9) by means of the following basic result.

6.1.1 (Maximum–Minimum Principle)   Assume that (10) holds. If

$$\Gamma_i \equiv a_i y_i - b_i y_{i-1} - c_i y_{i+1} \leq 0, \qquad i = 1, \ldots, n \tag{11}$$

then

$$y_i \leq \max(y_0, y_{n+1}), \qquad i = 1, \ldots, n. \tag{12}$$

If $\Gamma_i \geq 0$, $i = 1, \ldots, n$, then

$$y_i \geq \min(y_0, y_{n+1}), \qquad i = 1, \ldots, n. \tag{13}$$

Proof: Suppose that $\max_{0 \leq i \leq n+1} y_i$ is achieved at y_k for $1 \leq k \leq n$. Set $\mu = b_k/a_k$ and $\eta = c_k/a_k$. Then $\mu + \eta \leq 1$ and, by (11),

$$y_k = (1/a_k)[\Gamma_k + b_k y_{k-1} + c_k y_{k+1}] \leq \mu y_{k-1} + \eta y_{k+1} \leq \max(y_{k-1}, y_{k+1}).$$

But, by assumption, the last term is less than or equal to y_k and hence equality holds throughout. Suppose that $y_k = y_{k+1} > y_{k-1}$. Then, since $\mu > 0$,

$$y_k < \mu y_{k-1} + \eta y_{k+1} \leq y_k.$$

Therefore, it must be that $y_{k+1} = y_k = y_{k-1}$. Continuing this procedure at each point then shows that $y_0 = \cdots = y_{n+1}$. That is, if $\{y_i\}$ takes on a maximum at an interior point, it is constant. A similar argument proves the minimum result. $\$\$\$

As an immediate consequence of this result we obtain:

6.1.2   Assume that (10) holds. Then the system (9), with $y_0 = \alpha$, $y_{n+1} = \beta$, has a unique solution.

Proof: Write (9) in matrix form as $A\mathbf{y} = \mathbf{d}$ where

$$
A = \begin{bmatrix}
a_1 & -c_1 & & & \bigcirc \\
-b_2 & a_2 & \cdot & & \\
& \cdot & \cdot & \cdot & \\
& & \cdot & \cdot & -c_{n-1} \\
\bigcirc & & & -b_{n-1} & a_n
\end{bmatrix},
\qquad
d = \begin{bmatrix}
-r_1 h^2 + b_1 \alpha \\
-r_2 h^2 \\
\vdots \\
-r_{n-1} h^2 \\
-r_n h^2 + c_n \beta
\end{bmatrix}
\tag{14}
$$

and consider the homogeneous system $A\mathbf{y} = \mathbf{0}$. If we set $\alpha = \beta = 0$, and assume that $\Gamma_i = 0$, $i = 1, \ldots, n$, where Γ_i is defined by (11), we see that (12) and (13) imply that $y_i = 0$, $i = 1, \ldots, n$. That is, the system $A\mathbf{y} = \mathbf{0}$ has only the trivial solution and, hence, A is nonsingular. $\$\$\$

We next give the basic discretization error result. For simplicity, we shall assume that (5) is strengthened to

$$q(x) \geq \gamma > 0, \qquad x \in [a, b]. \tag{15}$$

This condition is by no means necessary,† however, and in the next section

---

† Condition (5), also, is only a convenient, and by no means necessary, restriction on q.

we shall return to (5). We also assume that (4) has a solution y and define the **local discretization error** by

$$\tau(x, h) = \frac{1}{h^2} [y(x - h) - 2y(x) + y(x + h)]$$

$$- \frac{p(x)}{2h} [y(x + h) - y(x - h)] - q(x)y(x) - r(x). \qquad (16)$$

6.1.3 (Gerschgorin's Theorem) Assume that (4) has a unique solution y and that (6) and (15) hold. Then there is a constant c, independent of h, such that

$$|y(x_i) - y_i(h)| \leq c\tau(h), \qquad i = 1, \ldots, n \qquad (17)$$

where $x_i = a + ih$, $y_1(h), \ldots, y_n(h)$ is the solution of (7) with $y_0 = \alpha$, $y_{n+1} = \beta$, and

$$\tau(h) = \max_{a+h \leq x \leq b-h} |\tau(x, h)| \qquad (18)$$

with $\tau(x, h)$ defined by (16).

Proof: For fixed h, set $e_i = y_i(h) - y(x_i)$, $i = 0, \ldots, n + 1$. Then from (9) and (16) we obtain

$$-b_i e_{i-1} + a_i e_i - c_i e_{i+1} = h^2 \tau(x_i, h), \qquad i = 1, \ldots, n$$

where a_i, b_i, c_i are given by (8). If we set $e = \max_{1 \leq i \leq n} |e_i|$, note that $e_0 = e_{n+1} = 0$, and use (6), we see that $|b_i| + |c_i| = b_i + c_i = 2$, and thus

$$|a_i e_i| \leq b_i |e_{i-1}| + c_i |e_{i+1}| + h^2 \tau(h) \leq 2e + h^2 \tau(h).$$

Now, by (15), $a_i \geq 2 + h^2 \gamma$ so that

$$(2 + h^2 \gamma)|e_i| \leq 2e + h^2 \tau(h), \qquad i = 1, \ldots, n$$

and hence

$$(2 + h^2 \gamma)e \leq 2e + h^2 \tau(h). \quad \$\$\$$$

The estimate (17) shows that the global discretization error,

$$\max_{1 \leq i \leq n} |y(x_i) - y_i(h)|$$

tends to zero as $h \to 0$ provided that $\lim_{h \to 0} \tau(h) = 0$. We next examine τ in more detail.

We first write $\tau(x, h)$ in the form

$$\tau(x, h) = \left\{\frac{1}{h^2} \left[y(x - h) - 2y(x) + y(x + h)\right] - y''(x)\right\}$$

$$+ p(x)\left\{\frac{1}{2h} \left[y(x - h) - y(x + h)\right] - y'(x)\right\} \tag{19}$$

by putting into (16) the value of $r(x)$ from the differential equation (4). We note, in particular, that τ is independent of q and r and depends essentially only on the finite difference approximations to y' and y''. It is easy to see (E6.1.3) that if y is twice continuously differentiable then

$$\lim_{h \to 0} \frac{1}{h^2} \left[y(x - h) - 2y(x) + y(x + h)\right] = y''(x)$$

$$\lim_{h \to 0} \frac{1}{2h} \left[y(x + h) - y(x - h)\right] = y'(x) \tag{20}$$

uniformly in x. Hence $\tau(h) \to 0$ as $h \to 0$ and as a first corollary to 6.1.3 we have:

6.1.4 If the solution of (4) is twice continuously differentiable and the conditions of **6.1.3** hold, then

$$\lim_{h \to 0} |y(x_i) - y_i(h)| = 0 \tag{21}$$

where h tends to zero so that $x_i = a + ih$ remains constant.

If we assume more regularity for the solution y, then we can obtain an estimate of the rate of convergence. In particular, if we assume that y is four times continuously differentiable and use the Taylor expansions

$$y(x \pm h) = y(x) \pm y'(x)h + \tfrac{1}{2}y''(x)h^2 \pm \tfrac{1}{6}y^{(3)}(x)h^3 + \tfrac{1}{24}y^{(4)}(\xi_\pm)h^4 \tag{22}$$

we obtain

$$\frac{1}{h^2} \left[y(x - h) - 2y(x) + y(x + h)\right] = y''(x) + \frac{1}{24} \left[y^{(4)}(\xi_+) + y^{(4)}(\xi_-)\right]h^2$$

and

$$\frac{1}{2h}\left[y(x + h) - y(x - h)\right] = y'(x) + \frac{1}{6} y^{(3)}(x)h^2 + \frac{h^3}{48}\left[y^{(4)}(\xi_+) - y^{(4)}\xi_-)\right].$$

Therefore, if we set

$$\omega = \max_{a \le x \le b} |p(x)|, \qquad M_i = \max_{a \le x \le b} |y^{(i)}(x)|, \qquad i = 3, 4$$

we obtain

$$\tau(h) = \max_{a+h \le x \le b-h} |\tau(x, h)| \le \tfrac{1}{12} M_4 h^2 + \tfrac{1}{6} M_3^2 \omega h^2 + \tfrac{1}{24} M_4 \omega h^3$$

$$\equiv c_1 h^2 + c_2 h^3. \tag{23}$$

If we put this estimate in (17), we can then summarize the rate of convergence result as follows.

6.1.5   Assume that the solution y of (4) is four times continuously differentiable and that the conditions of **6.1.3** hold. Then

$$|y(x_i) - y_i(h)| = O(h^2).$$

EXERCISES

E6.1.1   Let A be the tridiagonal matrix

$$A = \begin{bmatrix} 2 & -1 & & & \\ -1 & \cdot & & \ddots & \\ & & \ddots & & -1 \\ & & & -1 & 2 \end{bmatrix}.$$

Show that A has eigenvalues

$$\lambda_k = 2 - 2 \cos \frac{k\pi}{n+1}, \qquad k = 1, \ldots, n$$

and corresponding eigenvectors

$$\left(\sin \frac{k\pi}{n+1}, \sin \frac{2k\pi}{n+1}, \ldots, \sin \frac{nk\pi}{n+1} \right)^{\mathrm{T}}, \qquad k = 1, \ldots, n.$$

E6.1.2   Assume that the solution of the equations

(a) $y''(x) = y(x) + 1 - x^2$, $y(0) = 0, \quad y(1) = 1$
(b) $y''(x) = (2 + x)[y'(x) + y(x)] + x$, $y(0) = 0, \quad y(1) = 1$

exist and are four times continuously differentiable with $|y^{(3)}(x)| \leq M_3$, $|y^{(4)}(x)| \leq M_4$. Find the largest h_0 such that you can guarantee that for all $0 < h \leq h_0$, the discretization error satisfies $|y(x_i) - y_i(h)| \leq 10^{-6}$, $i = 1, \ldots, n$.

E6.1.3   Assume that y is twice continuously differentiable. Show that (20) holds.

6.2 MATRIX METHODS

We turn next to a discussion of the discrete boundary value problem of the previous section by means of matrix theoretic methods. We shall introduce several important classes of matrices that arise naturally in connection with boundary value problems. Moreover, many of the results we prove here will also be of value in our study of iterative methods in the next chapter.

In the previous section, we showed that the matrix

$$
A = \begin{bmatrix}
a_1 & -c_1 & & \text{\Large O} \\
-b_2 & \cdot & & \ddots \\
& & \ddots & & -c_{n-1} \\
\text{\Large O} & & -b_n & a_n
\end{bmatrix}
\tag{1}
$$

where

$$
a_1 > c_1 > 0, \quad a_n > b_n > 0, \quad b_i > 0, \quad c_i > 0, \quad a_i \geq b_i + c_i,
$$
$$
i = 2, \ldots, n-1 \tag{2}
$$

is nonsingular, by means of the maximum principle. We now establish this result as a consequence of certain interesting results in matrix theory.

Recall that a **permutation matrix** P is defined by the property that each row and column of P has exactly one element equal to one while the other elements are zero. Some easily verified properties of permutation matrices are given in E6.2.1.

6.2.1   Definition   An $n \times n$ real or complex matrix A is **reducible** if there is a permutation matrix P so that

$$
PAP^{-1} = \begin{bmatrix} B_{11} & B_{12} \\ 0 & B_{22} \end{bmatrix}
$$

where B_{11} and B_{22} are square matrices. A matrix A is **irreducible** if it is not reducible.

Clearly, any matrix all of whose elements are nonzero is irreducible. More generally, $A \in L(C^n)$ is reducible if and only if there is a nonempty subset of indices $J \subset \{1, \ldots, n\}$ such that

$$a_{kj} = 0, \qquad k \in J, \quad j \notin J. \tag{3}$$

For example, the matrix

$$A = \begin{bmatrix} 1 & 0 & 1 \\ 0 & 1 & 0 \\ 1 & 1 & 1 \end{bmatrix}$$

is reducible since with the permutation P which interchanges the second and third rows

$$PAP^{-1} = \begin{bmatrix} 1 & 1 & 0 \\ 1 & 1 & 1 \\ 0 & 0 & 1 \end{bmatrix}.$$

Equivalently, we may take the set J of (3) to be $\{2\}$.

The criterion (3) is simply a restatement of the definition 6.2.1. A less obvious equivalency is the following.

6.2.2 A matrix $A \in L(C^n)$ is irreducible if and only if for any two distinct indices $1 \le i, j \le n$, there is a sequence of nonzero elements of A of the form

$$\{a_{i,i_1}, a_{i_1 i_2}, \ldots, a_{i_m j}\}. \tag{4}$$

Proof: If there is a sequence of nonzero elements of the form (4), then we will say that there is a **chain** for i, j. We first prove the sufficiency by contradiction. That is, suppose that a chain exists for all i, j but that A is reducible. Let J be such that (3) holds and choose $i \in J, j \notin J$. By assumption, there is a chain (4) and $a_{i, i_1} \ne 0$. Hence $i_1 \in J$ and since $a_{i_1 i_2} \ne 0$ also $i_2 \in J$. Continuing in this way we see that $i_m \in J$. But $j \notin J$ and hence $a_{i_m j} = 0$ so that the chain cannot be completed.

Conversely, assume that A is irreducible and, for given i, set $J = \{k : a$ chain exists for $i, k\}$. Clearly, J is not empty since, otherwise, $a_{ik} = 0$, $k = 1, \ldots, n$, and this would contradict the irreducibility. Now suppose that, for some j, there is no chain for i, j. Then J is not the entire set $\{1, \ldots, n\}$ and we claim that

$$a_{kl} = 0, \qquad k \in J, \quad l \notin J \tag{5}$$

which would contradict the irreducibility of A. But (5) follows immediately from the fact that there is a chain for i, k so that if $a_{kl} \neq 0$, we may add a_{kl} to obtain a chain for i, l. This implies that $l \in J$. $\$\$\$$

As an application of this result, we have the following.

6.2.3   Assume that $b_2, \ldots, b_n$ and $c_1, \ldots, c_{n-1}$ are all nonzero. Then the matrix (1) is irreducible.

Proof: If $1 \leq i < j \leq n$, then the elements $c_i, \ldots, c_{j-1}$ satisfy the conditions of **6.2.2**, while if $1 \leq j < i \leq n$ we may take $b_{j+1}, \ldots, b_i$. $\$\$\$$

It is useful to examine the previous result in a more geometric way. With any $n \times n$ matrix A, we associate a **graph** as follows. We consider n distinct points $P_1, \ldots, P_n$, which we call **nodes**, and for every nonzero element a_{ij} of A we construct a **directed path** or **directed link** from P_i to P_j:

$$P_i \qquad P_j.$$

As a result, we associate with A a **directed graph**. For example, consider the matrix

$$A = \begin{bmatrix} 0 & 0 & 1 & 1 \\ 0 & 0 & 1 & 1 \\ 1 & 0 & 1 & 0 \\ 1 & 1 & 0 & 1 \end{bmatrix};$$

then the corresponding directed graph is shown in Figure 6.2.1.

We say that a directed graph is **strongly connected** if for any pair of nodes P_i, P_j there is a path

$$\overrightarrow{P_i P_{i_1}}, \overrightarrow{P_{i_1} P_{i_2}}, \ldots, \overrightarrow{P_{i_m} P_j}$$

connecting P_i and P_j. By comparison of this definition with **6.2.2**, it is clear that we have the following characterization of irreducibility.

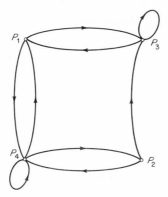

Figure 6.2.1

6.2.4 A matrix $A \in L(C^n)$ is irreducible if and only if its associated directed graph is strongly connected.

As another example, we construct the associated graph for the matrix (1) under the assumption that $c_1, \ldots, c_{n-1}$ and $b_2, \ldots, b_n$ are all nonzero

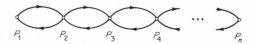

Figure 6.2.2

(see Figure 6.2.2). Clearly, this graph is strongly connected, and we conclude again the result **6.2.3**.

We continue now with some invertibility theorems. We first need to delineate two important classes of matrices.

6.2.5 **Definition**   A matrix $A \in L(C^n)$ is **diagonally dominant** if

$$|a_{ii}| \geq \sum_{j \neq i} |a_{ij}|, \qquad i = 1, \ldots, n \tag{6}$$

and **strictly diagonally dominant** if strict inequality holds in (6) for all i. The matrix is **irreducibly diagonally dominant** if it is irreducible, diagonally dominant, and strict inequality holds in (6) for at least one i.

6.2.6 (Diagonal Dominance Theorem) Assume that $A \in L(C^n)$ is either strictly or irreducibly diagonally dominant. Then A is nonsingular.

Proof: Assume first that A is strictly diagonally dominant and suppose that there is an $\mathbf{x} \neq 0$ such that $A\mathbf{x} = 0$. Let

$$|x_k| = \max_{1 \leq j \leq n} |x_j|.$$

Then $|x_k| > 0$, and the inequality

$$|a_{kk}| |x_k| = \left| \sum_{j \neq k} a_{kj} x_j \right| \leq |x_k| \sum_{j \neq k} |a_{kj}| \tag{7}$$

contradicts the strict diagonal dominance. Similarly, if A is irreducibly diagonally dominant, again suppose that there is an $\mathbf{x} \neq 0$ so that $A\mathbf{x} = 0$, and let m be such that

$$|a_{mm}| > \sum_{j \neq m} |a_{mj}|. \tag{8}$$

Next, define the set of indices

$$J = \{k : |x_k| \geq |x_i|, \quad i = 1, \ldots, n; \quad |x_k| > |x_j| \quad \text{for some } j\}.$$

Clearly, J is not empty, for this would imply that $|x_1| = \cdots = |x_n| \neq 0$ and hence the inequality (7), with $k = m$, would contradict (8). Now, for any $k \in J$,

$$|a_{kk}| \leq \sum_{j \neq k} |a_{kj}| |x_j| / |x_k|$$

and it follows that $a_{kj} = 0$ whenever $|x_k| > |x_j|$, or else the diagonal dominance is contradicted. But for any $k \in J$ and $j \notin J$, we must have $|x_k| > |x_j|$ and therefore

$$a_{kj} = 0, \qquad k \in J, \quad j \notin J.$$

But then A is reducible, and this is a contradiction. $$$

As an example of the use of this theorem we have:

6.2.7 Assume that the matrix (1) satisfies (2). Then A is nonsingular.

The proof is immediate since we have shown in 6.2.3 that A is irreducible, and (2) shows that A is diagonally dominant as well as $a_1 > c_1$. Hence A is irreducibly diagonally dominant.

Another corollary of interest is the following.

6.2.8 If $A \in L(R^n)$ is symmetric, irreducibly diagonally dominant, and has positive diagonal elements, then A is positive definite. In particular, if the matrix A of (1) satisfies (2) and is symmetric, then it is positive definite.

Proof: Since A is symmetric, its eigenvalues λ_i are real and the Gerschgorin circle theorem 3.2.1 shows, since A is diagonally dominant, that $\lambda_i \geq 0$, $i = 1, \ldots, n$. But by 6.2.6 A is nonsingular so that $\lambda_i > 0$, $i = 1, \ldots, n$. $\$\$\$

We turn next to some results which involve the idea of a partial ordering. For vectors $\mathbf{x}, \mathbf{y} \in R^n$ we define the **natural** or **componentwise partial ordering** by

$$\mathbf{x} \leq \mathbf{y} \qquad \text{if and only if} \qquad x_i \leq y_i, \quad i = 1, \ldots, n$$

and similarly for matrices $A, B \in L(R^n)$,

$$A \leq B \qquad \text{if and only if} \qquad a_{ij} \leq b_{ij}, \quad i, j = 1, \ldots, n.$$

In the special case that $\mathbf{x} = \mathbf{0}$ or $A = 0$ we say that $\mathbf{y} \geq \mathbf{0}$ or $B \geq 0$ is **nonnegative**. An important class of nonnegative vectors or matrices is given by the **absolute values**:

$$|\mathbf{x}| = (|x_1|, \ldots, |x_n|)^T, \qquad |A| = (|a_{ij}|).$$

We shall be interested primarily in conditions which ensure that a given matrix has a nonnegative inverse. Our first result follows from the Neumann lemma 1.3.10.

6.2.9 Let $B \in L(R^n)$ be nonnegative. Then $(I - B)^{-1}$ exists and is nonnegative if and only if $\rho(B) < 1$.

Proof: If $\rho(B) < 1$, then 1.3.10 ensures that $(I - B)^{-1}$ exists and that $(I - B)^{-1} = \sum_0^\infty B^i$. Since every member of this sum is nonnegative, it follows that $(I - B)^{-1} \geq 0$. Conversely, assume that $(I - B)^{-1} \geq 0$ and let λ be any eigenvalue of B with corresponding eigenvector $\mathbf{x} \neq \mathbf{0}$. Then $|\lambda| \, |\mathbf{x}| \leq B|\mathbf{x}|$ so that $(I - B)|\mathbf{x}| \leq (1 - |\lambda|)|\mathbf{x}|$. Hence

$$|\mathbf{x}| \leq (1 - |\lambda|)(I - B)^{-1}|\mathbf{x}|$$

and, since $\mathbf{x} \neq \mathbf{0}$, it follows that $|\lambda| < 1$; that is, $\rho(B) < 1$. $\$\$\$

A particularly important class of matrices with nonnegative inverses is given by the following:

6.2.10   Definition   A matrix $A \in L(R^n)$ is an **M-matrix** if $A^{-1} \geq 0$ and $a_{ij} \leq 0$, $i \neq j$. A symmetric M-matrix is a **Stieltjes** matrix.

The next result gives a characterization of M-matrices.

6.2.11   Let $A \in L(R^n)$ satisfy $a_{ij} \leq 0$, $i \neq j$. Then A is an M-matrix if and only if $a_{ii} > 0$, $i = 1, \ldots, n$, and the matrix $B = I - D^{-1}A$, where $D = \mathrm{diag}(a_{11}, \ldots, a_{nn})$, satisfies $\rho(B) < 1$.

Proof: Suppose that $\rho(B) < 1$ and that $a_{ii} > 0$, $i = 1, \ldots, n$. Then, since $B \geq 0$, 6.2.9 ensures that $(D^{-1}A)^{-1} = (I - B)^{-1} \geq 0$, so that A^{-1} exists and, since $D \geq 0$, $A^{-1} \geq 0$. Conversely, if A is an M-matrix, then the diagonal elements of A are positive because if some $a_{ii} < 0$, then $\mathbf{a}_i \leq \mathbf{0}$, where $\mathbf{a}_i$ is the ith column of A. Hence, $\mathbf{e}_i = A^{-1}\mathbf{a}_i \leq 0$, where $\mathbf{e}_i$ is the ith coordinate vector, and this is a contradiction. Hence $D \geq 0$, and D^{-1} exists. It follows that $B \geq 0$ and $(I - B)^{-1} = A^{-1}D \geq 0$ so that 6.2.9 ensures that $\rho(B) < 1$.   \$\$\$

We next give a basic comparison result for spectral radii.

6.2.12   Let $B \in L(R^n)$ and $C \in L(C^n)$. If $|C| \leq B$, then $\rho(C) \leq \rho(B)$.

Proof: Set $\sigma = \rho(B)$, let $\varepsilon > 0$ be arbitrary, and set
$$B_1 = (\sigma + \varepsilon)^{-1}B, \qquad C_1 = (\sigma + \varepsilon)^{-1}C.$$
Clearly,
$$|C_1|^k \leq B_1^k, \qquad k = 0, 1, \ldots.$$
But $\rho(B_1) < 1$ so that $B_1^k \to 0$ as $k \to \infty$ and hence $C_1^k \to 0$ as $k \to \infty$. But then $\rho(C_1) < 1$. Therefore $\rho(C) < \sigma + \varepsilon$ and, since ε was arbitrary, $\rho(C) \leq \sigma$.   \$\$\$

As an immediate corollary, we have the following useful result.

6.2.13 Let $C \in L(R^n)$. Then $\rho(C) \le \rho(|C|)$.

It is not in general true that the sum of M-matrices is again an M-matrix (E6.2.8). However, a special case of importance is the following.

6.2.14 Let $A \in L(R^n)$ be an M-matrix and $D \in L(R^n)$ be a nonnegative diagonal matrix. Then $A + D$ is an M-matrix and $(A + D)^{-1} \le A^{-1}$.

Proof: Write $A = D_1 - H$ where $D_1 = \text{diag}(a_{11}, \ldots, a_{nn})$ and set $B_1 = D_1^{-1}H$, $B = (D + D_1)^{-1}H$. Since A is an M-matrix, we have $H \ge 0$, $D_1 \ge 0$, and, clearly, $0 \le B \le B_1$. But **6.2.11** shows that $\rho(B_1) < 1$ so that, by **6.2.12**, $\rho(B) < 1$. Hence **6.2.11** ensures that $A + D$ is an M-matrix. The inequality follows by multiplying $A \le A + D$ by A^{-1} and $(A + D)^{-1}$. $\$\$\$$

As an immediate corollary of this theorem we obtain the following result on symmetric M-matrices.

6.2.15 Let $A \in L(R^n)$ be a Stieltjes matrix. Then A is positive definite.

Proof: Suppose that A has an eigenvalue $\lambda \le 0$. Then **6.2.14** shows that $A - \lambda I$ is an M-matrix and hence nonsingular. But this is a contradiction since $A - \lambda I$ is singular if λ is an eigenvalue. $\$\$\$$

The characterization theorem **6.2.11** is not very useful for checking whether a given matrix is an M-matrix. We next give a useful sufficient condition which is a consequence of the following result.

6.2.16 Let $B \in L(C^n)$ be irreducible. If

$$\sum_{j=1}^{n} |b_{ij}| \le 1, \qquad i = 1, \ldots, n \tag{9}$$

and strict inequality holds for at least one i, then $\rho(B) < 1$.

Proof: Clearly (9) implies that $\rho(B) \le \|B\|_\infty \le 1$. Now suppose that $\rho(B) = 1$ and let λ be any eigenvalue of modulus unity. Then $\lambda I - B$ is singular, but, by (9),

$$|\lambda - b_{ii}| \ge 1 - |b_{ii}| \ge \sum_{j \ne i} |b_{ij}|$$

and strict inequality holds for at least one i. Therefore, since $\lambda I - B$ is irreducible with B, 6.2.6 shows that $\lambda I - B$ is nonsingular. This is a contradiction and we must have $\rho(B) < 1$. $\$\$\$$

6.2.17   Let $A \in L(R^n)$ be strictly or irreducibly diagonally dominant and assume that $a_{ij} \leq 0$, $i \neq j$, and $a_{ii} > 0$, $i = 1, \ldots, n$. Then A is an M-matrix.

Proof: Define $B \in L(R^n)$ by $B = I - D^{-1}A$ where, again, D is the diagonal part of A. Then, by 6.2.11, it suffices to show that $\rho(B) < 1$. By the diagonal dominance of A, (9) holds for B and if A is strictly diagonally dominant, then strict inequality holds in (9) for all i. In this case $\rho(B) \leq \|B\|_\infty < 1$. On the other hand, if A is irreducibly diagonally dominant, then the result follows immediately from 6.2.16. $\$\$\$$

For a corollary, we consider the tridiagonal matrix (1).

6.2.18   Let the matrix (1) satisfy (2). Then A is an M-matrix.

The proof is an immediate consequence of the previous result since (2) ensures that A is irreducibly diagonally dominant.

To complete this section, we prove another version of Gerschgorin's theorem for the boundary value problem

$$y''(x) = q(x)y(x) + r(x), \qquad y(a) = \alpha, \quad y(b) = \beta \tag{10}$$

under the condition

$$q(x) \geq 0, \qquad x \in [a, b]. \tag{11}$$

Since, for simplicity, we have set $p(x) \equiv 0$, the difference equations of the previous section now take the simpler form

$$y_{i-1} - 2y_i + y_{i+1} = h^2 q_i y_i + h^2 r_i, \qquad i = 1, \ldots, n, \quad y_0 = \alpha, \quad y_{n+1} = \beta \tag{12}$$

with $q_i = q(x_i)$ and $r_i = r(x_i)$. These equations we can write as

$$A\mathbf{y} = \mathbf{b} \tag{13}$$

where

$$
A = \begin{bmatrix}
2 + h^2 q_1 & -1 & & & \\
-1 & 2 + h^2 q_2 & -1 & & \\
& -1 & \ddots & \ddots & \\
& & \ddots & \ddots & -1 \\
& & & -1 & 2 + h^2 q_n
\end{bmatrix}, \quad
b = \begin{bmatrix}
-h^2 r_1 + \alpha \\
-h^2 r_2 \\
\vdots \\
-h^2 r_{n-1} \\
-h^2 r_n + \beta
\end{bmatrix}
$$

(14)

We will now prove the following complement of 6.1.3.

6.2.19 (Gerschgorin's Theorem)   Assume that (10) has a solution y and that (11) holds. Then there is a constant $c > 0$ such that

$$
|y(x_i) - y_i(h)| \le c\tau(h), \qquad i = 1, \ldots, n
$$

(15)

where $y_i(h)$, $i = 1, \ldots, n$ is the unique solution of (12), and

$$
\tau(h) = \max_{a + h \le x \le b - h} |\tau(x, h)|
$$

with

$$
\tau(x, h) = \frac{1}{h^2} [y(x - h) - 2y(x) + y(x + h)] - q(x)y(x) - r(x).
$$

(16)

Proof: Let $\hat{y} = (y(x_1), \ldots, y(x_n))^T$ and $\Gamma = (\tau(x_1, h), \ldots, \tau(x_n, h))^T$. Then, by (16),

$$
A\hat{y} = \mathbf{b} - h^2 \Gamma
$$

so that subtracting this from (13) gives

$$
A(\mathbf{y} - \hat{y}) = h^2 \Gamma.
$$

Now let A_0 be the matrix obtained from A when $q(x) \equiv 0$. Then both A_0 and A are irreducibly diagonally dominant and hence, by 6.2.6, nonsingular. Moreover, both A_0 and A are M-matrices by 6.2.18, and, by 6.2.14, $A^{-1} \le A_0^{-1}$. Hence, using the absolute value, we have

$$
|\mathbf{y} - \hat{y}| = h^2 |A^{-1}\Gamma| \le h^2 A^{-1} |\Gamma| \le h^2 A_0^{-1} |\Gamma| \le h^2 \tau(h) A_0^{-1}\mathbf{e}
$$

(17)

where $\mathbf{e} = (1, 1, \ldots, 1)^T$. We therefore need to evaluate $A_0^{-1}\mathbf{e}$. Let

$$
w_i = \tfrac{1}{2}(x_i - a)(b - x_i), \qquad i = 0, \ldots, n + 1.
$$

Then $w_0 = w_{n+1} = 0$, and it is easy to verify that

$$-w_{i-1} + 2w_i - w_{i+1} = h^2, \qquad i = 1, \ldots, n$$

which is equivalent to $A_0 \mathbf{w} = h^2 \mathbf{e}$, with $\mathbf{w} = (w_1, \ldots, w_n)^T$. Hence $A_0^{-1} \mathbf{e} = h^{-2} \mathbf{w}$ so that (17) becomes

$$|\mathbf{y} - \hat{\mathbf{y}}| \le \tau(h) |\mathbf{w}|$$

or

$$|y_i(h) - y(x_i)| \le \tau(h) \max_{1 \le i \le n} |w_i| \le \tfrac{1}{8}(b - a)^2 \tau(h). \quad \$\$\$$$

The local truncation error $\tau(h)$ may be estimated as in the previous section. Hence, corresponding to 6.1.5, we have the following corollary.

6.2.20 Assume that the solution y of (10) is four times continuously differentiable and that the conditions of 6.2.19 hold. Then

$$|y(x_i) - y_i(h)| = O(h^2).$$

EXERCISES

E6.2.1   Let $P \in L(R^n)$ be a permutation matrix. Show that

(a) P is orthogonal
(b) P is obtained from the identity matrix by permutation of rows or columns
(c)   If $A \in L(R^n)$, the effect of the multiplication PA is to permute the rows of A while the multiplication AP permutes the columns of A.

E6.2.2   Decide whether each of the following matrices is reducible or irreducible:

(a) $\begin{bmatrix} 2 & 1 & 1 \\ 0 & 1 & 1 \\ 1 & 0 & 1 \end{bmatrix}$ (b) $\begin{bmatrix} 0 & 1 & 0 \\ 1 & 0 & 1 \\ 0 & 1 & 0 \end{bmatrix}$ (c) $\begin{bmatrix} 0 & 0 & 1 \\ 0 & 1 & 0 \\ 1 & 0 & 0 \end{bmatrix}$

E6.2.3 Prove the Gerschgorin circle theorem 3.2.1 as a corollary of 6.2.6. Prove that a strictly diagonally dominant matrix is nonsingular as a corollary of 3.2.1.

E6.2.4 Which of the matrices of E6.2.2 are diagonally dominant? strictly diagonally dominant? irreducibly diagonally dominant?

E6.2.5 Let $A, B \in L(R^n)$ be the tridiagonal matrices

$$A = \begin{bmatrix} a_1 & b_1 & & & \bigcirc \\ c_1 & \cdot & \cdot & & \\ & \cdot & \cdot & \cdot & \\ & & \cdot & \cdot & b_{n-1} \\ \bigcirc & & & c_{n-1} & a_n \end{bmatrix}, \quad B = \begin{bmatrix} a_1 & \gamma_1 & & & \bigcirc \\ \gamma_1 & \cdot & \cdot & & \\ & \cdot & \cdot & \cdot & \\ & & \cdot & \cdot & \gamma_{n-1} \\ \bigcirc & & & \gamma_{n-1} & a_n \end{bmatrix}$$

where $b_i c_i > 0$ and $\gamma_i = \sqrt{b_i c_i}$, $i = 1, \ldots, n-1$. Show that if

$$D = \text{diag}\left(1, \frac{b_1}{c_1}, \frac{b_1 b_2}{c_1 c_2}, \ldots, \frac{b_1 \cdots b_{n-1}}{c_1 \cdots c_{n-1}}\right)$$

then $B = DAD^{-1}$. Show also that if $a_i \geq |b_i| + |c_{i-1}|$, $i = 2, \ldots, n-1$ and $a_1 > |b_1|$, $a_n > |c_{n-1}|$, then B is positive definite.

E.6.2.6 Let $A \in L(R^n)$. Show that $A^{-1} \geq 0$ if and only if there exist non-singular, nonnegative $P, Q \in L(R^n)$ such that $PAQ = I$.

E6.2.7 Which of the matrices of E6.2.2 are M-matrices? Stieltjes matrices?

E6.2.8 Give an example of a 2×2 M-matrix which is neither irreducible nor diagonally dominant. Give an example of two 2×2 M-matrices whose sum is not an M-matrix.

E6.2.9 Try an alternative proof of 6.2.19 along the following lines. Use 1.3.3 to show that $\lambda_0 \leq \lambda$, where λ_0 and λ are the lowest eigenvalues of A_0 and A, respectively. Hence conclude that $\|A^{-1}\|_2 \leq \|A_0^{-1}\|_2$. Now use E6.1.1 to compute $\|A_0^{-1}\|_2$ explicitly and then modify the proof of 6.2.19 accordingly. How does your bound compare with (15)?

E6.2.10 Prove Theorem 6.1.3 along the lines of 6.2.19 as follows: Show that $A(\mathbf{y} - \hat{\mathbf{y}}) = h^2 \mathbf{\Gamma}$, where A is given by (14), and then prove that $\|A^{-1}\|_\infty \leq \gamma^{-1} h^2$ by showing that $\|A\mathbf{z}\|_\infty \geq \gamma h^2 \|\mathbf{z}\|_\infty$ for all $\mathbf{z} \in R^n$.

READING

The treatment of Section 6.1 follows to some degree that of Isaacson and Keller [1966], while that of Section 6.2 is along the lines pursued in Varga [1962] and Ortega and Rheinboldt [1970]. For additional results on discretization error for two-point boundary value problems, including nonlinear problems, see Henrici [1962] and Keller [1968]. For extension to partial differential equations, see Forsythe and Wasow [1960]. Excellent references for the matrix theory of Section 6.2 are Householder [1964] and Varga [1962].

PART III

CONVERGENCE OF ITERATIVE METHODS

In the previous part we studied the discretization error that arises when a "continuous problem" is replaced by a discrete one. We now take up the next basic error—convergence error—which results when we truncate an infinite sequence to finitely many terms. There are many intrinsic similarities between this type of error and discretization error, but the spirit in which we approach the errors is entirely different. There are also many different types of problems in which convergence error may arise: for example, summing an infinite series and analyzing the convergence of sequences generated in many different ways. In the next two chapters we shall restrict our attention to those sequences generated by iterative processes.

CHAPTER 7

SYSTEMS OF LINEAR EQUATIONS

7.I CONVERGENCE

Consider the system of linear equations

$$A\mathbf{x} = \mathbf{b} \tag{1}$$

where $A \in L(R^n)$ is assumed to be nonsingular. If n is reasonably small (say, $n \leq 200$), the method of Gaussian elimination, to be discussed in Chapter 9, is probably the most efficient method to approximate a solution of (1). However, for certain problems arising, for example, from differential equations, n may be very large (say, $n \geq 10^4$), and it may be desirable to use an iterative method for (1).

One of the simplest iterative methods is that of **Jacobi**. Assume that

$$a_{ii} \neq 0, \qquad i = 1, \ldots, n \tag{2}$$

and that the kth vector iterate $\mathbf{x}^k$ has been computed; then the components of the next iterate $\mathbf{x}^{k+1}$ are given by

$$x_i^{k+1} = \frac{1}{a_{ii}} \left(b_i - \sum_{j \neq i} a_{ij} x_j^k \right), \qquad i = 1, \ldots, n. \tag{3}$$

If we let $D = \mathrm{diag}(a_{11}, \ldots, a_{nn})$ and $B = D - A$ then, clearly, (3) may be written as

$$\mathbf{x}^{k+1} = D^{-1}B\mathbf{x}^k + D^{-1}\mathbf{b}, \qquad k = 0, 1, \ldots. \tag{4}$$

A closely related iteration may be derived from the following observation. If we assume that the computations of (3) are done sequentially for $i = 1, 2, \ldots, n$, then at the time we are ready to compute x_i^{k+1} the new components $x_1^{k+1}, \ldots, x_{i-1}^{k+1}$ are available, and it would seem reasonable to use them instead of the old components; that is, we compute

$$x_i^{k+1} = \frac{1}{a_{ii}} \left(b_i - \sum_{j=1}^{i-1} a_{ij} x_j^{k+1} - \sum_{j=i+1}^{n} a_{ij} x_j^k \right), \qquad i = 1, \ldots, n. \quad (5)$$

This is the **Gauss–Seidel** iteration. Let $-L$ be the strictly lower triangular part of A and $-U$ the strictly upper triangular part; that is,

$$L = -\begin{bmatrix} 0 & & & \bigcirc \\ a_{21} & & \cdot & \\ \vdots & & & \cdot \\ a_{n1} & \cdots & a_{nn-1} & 0 \end{bmatrix}, \qquad U = -\begin{bmatrix} 0 & a_{12} & \cdots & a_{1n} \\ & \cdot & & \vdots \\ \bigcirc & & \cdot & a_{n-1,n} \\ & & & 0 \end{bmatrix}.$$

$$(6)$$

Then, with $D = \mathrm{diag}(a_{11}, \ldots, a_{nn})$, we have

$$A = D - L - U \quad (7)$$

and (5) is equivalent to

$$Dx^{k+1} = b + Lx^{k+1} + Ux^k.$$

The condition (2) ensures that $D - L$ is nonsingular and hence the Gauss–Seidel iteration may be written in the form†

$$x^{k+1} = (D - L)^{-1} U x^k + (D - L)^{-1} b, \qquad k = 0, 1, \ldots. \quad (8)$$

Both of the iterations (4) and (8) are of the general form

$$x^{k+1} = Hx^k + d, \qquad k = 0, 1, \ldots \quad (9)$$

and it is easy to see that $x^* = Hx^* + d$ if and only if $Ax^* = b$. We next prove the basic convergence theorem for the general iteration (9).

7.1.1   (Fundamental Theorem of Linear Iterative Methods)   Let $H \in L(R^n)$ and assume that the equation $x = Hx + d$ has a unique solution x^*. Then the iterates (9) converge to x^* for any x^0 if and only if $\rho(H) < 1$.

Proof: If we subtract $x^* = Hx^* + d$ from (9) we obtain the error equation

$$x^{k+1} - x^* = H(x^k - x^*) = \cdots = H^{k+1}(x^0 - x^*).$$

† But (3) and (5) should be used for actual computation, not (4) and (8).

Hence, in order that $\lim_{k\to\infty}(\mathbf{x}^k - \mathbf{x}^*) = \mathbf{0}$ for any $\mathbf{x}^0$ it is necessary and sufficient that $\lim_{k\to\infty}H^k = 0$. The result now follows from 1.3.9. $\$\$\$

It is interesting to note that we have previously proved 7.1.1 in the context of difference equations. Indeed, in Theorem 4.2.2 we showed that a solution of (9), considered to be a difference equation, is asymptotically stable if and only if $\rho(H) < 1$, and this result is precisely equivalent to the convergence theorem 7.1.1. We will see various other connections between the theories of difference equations and iterative methods in the sequel.

Theorem 7.1.1 reduces the convergence analysis of (9) to the algebraic problem of showing that $\rho(H) < 1$. We next give some sufficient conditions that this spectral radius condition be satisfied. In the following results, we use the notation for nonnegative matrices introduced in 6.2.

7.1.2 Definition Let $A, B, C \in L(R^n)$. Then $A = B - C$ is a **regular splitting** of A if $C \geq 0$ and B is nonsingular with $B^{-1} \geq 0$.

7.1.3 (Regular Splitting Theorem) Assume that $A \in L(R^n)$ is nonsingular, that $A^{-1} \geq 0$, and that $A = B - C$ is a regular splitting. Then $\rho(B^{-1}C) < 1$. Hence the iterates

$$\mathbf{x}^{k+1} = B^{-1}C\mathbf{x}^k + B^{-1}\mathbf{b}, \qquad k = 0, 1, \ldots$$

converge to $A^{-1}\mathbf{b}$ for any $\mathbf{x}^0$.

Proof: Set $H = B^{-1}C$. Then $H \geq 0$ and by the relations

$$(I + H + \cdots + H^m)(I - H) = I - H^{m+1}, \qquad B^{-1} = (I - H)A^{-1}$$

we have, since $A^{-1} \geq 0$,

$$0 \leq (I + H + \cdots + H^m)B^{-1} = (I - H^{m+1})A^{-1} \leq A^{-1}$$

for all $m \geq 0$. Since $B^{-1} \geq 0$, each row of B^{-1} must contain at least one positive element, and it follows that the elements of $I + \cdots + H^m$ are bounded above as $m \to \infty$. Therefore, since $H \geq 0$, the sum converges and, consequently, $\lim_{k\to\infty} H^k = 0$. Theorem 1.3.9 then shows that $\rho(H) < 1$. $\$\$\$

As an immediate corollary of this result, we have the following convergence theorem for the Jacobi and Gauss–Seidel iterations applied to an M-matrix (6.2.10).

7.1.4   Let $A \in L(R^n)$ be an M-matrix and let $\mathbf{b} \in R^n$ be arbitrary. Then the Jacobi iterates (4) and the Gauss–Seidel iterates (8) converge to $A^{-1}\mathbf{b}$ for any $\mathbf{x}^0$.

Proof: Since A is an M-matrix, it is nonsingular and $A^{-1} \geq 0$. Moreover, the matrices L and U of (6) are nonnegative and, by 6.2.11, $D = \text{diag}(a_{11}, \ldots, a_{nn})$ is nonnegative and nonsingular. Hence†

$$(D - L)^{-1} = [I + D^{-1}L + (D^{-1}L)^2 + \cdots + (D^{-1}L)^n]D^{-1} \geq 0 \qquad (10)$$

where the series expansion terminates because L is strictly lower triangular. Therefore $A = (D - L) - U$ is a regular splitting and clearly $A = D = (L + U)$ is also. Hence 7.1.3 applies. $\$\$\$$

We next give another convergence theorem which does not rely on the signs of the elements of A.

7.1.5   (Diagonal Dominance Theorem)   Assume that $\mathbf{b} \in R^n$ is arbitrary and that $A \in L(R^n)$ is either strictly or irreducibly diagonally dominant. Then both the Jacobi iterates (4) and the Gauss–Seidel iterates (8) converge to $A^{-1}\mathbf{b}$ for any $\mathbf{x}^0$.

Proof: As in **6.2**, let $|B|$ denote the matrix whose elements are the absolute values of those of B. Clearly, the matrix

$$\hat{A} = |D| - |L| - |U|$$

is strictly or irreducibly diagonally dominant with A. Hence, by 6.2.17, $\hat{A}$ is an M-matrix and therefore we have shown in 7.1.4 that

$$\rho\{|D|^{-1}(|L| + |U|)\} < 1, \qquad \rho\{(|D| - |L|)^{-1}|U|\} < 1.$$

But

$$|D^{-1}(L + U)| \leq |D|^{-1}(|L| + |U|)$$
$$|(D - L)^{-1}U| \leq (|D| - |L|)^{-1}|U|$$

where the second inequality follows from (10). Thus, by the comparison theorem 6.2.12,

$$\rho\{D^{-1}(L + U)\} \leq \rho\{|D|^{-1}(|L| + |U|)\} < 1$$

---

† This also follows from 6.2.9.

and

$$\rho\{(D - L)^{-1}U\} \le \rho\{(|D| - |L|)^{-1}|U|\} < 1.$$

The result then follows from 7.1.3. $$$

We turn next to a modification of the Gauss–Seidel iteration known as **successive overrelaxation** (SOR). In this iteration, the Gauss–Seidel iterate is computed as before by

$$\bar{x}_i^{k+1} = \frac{1}{a_{ii}} \left(b_i - \sum_{j=1}^{i-1} a_{ij} x_j^{k+1} - \sum_{j=i+1}^{n} a_{ij} x_j^k \right) \tag{11}$$

but the new value of x_i is taken to be

$$x_i^{k+1} = x_i^k + \omega(\bar{x}_i^{k+1} - x_i^k) \tag{12}$$

for some parameter ω. If $\omega = 1$, then clearly x_i^{k+1} is just the Gauss–Seidel iterate (5).

In order to write this procedure in matrix form, first substitute (11) into (12) to give

$$x_i^{k+1} = (1 - \omega)x_i^k + \frac{\omega}{a_{ii}} \left(b_i - \sum_{j=1}^{i-1} a_{ij} x_j^{k+1} - \sum_{j=i+1}^{n} a_{ij} x_j^k \right)$$

and then rearrange into the form

$$a_{ii} x_i^{k+1} + \omega \sum_{j=1}^{i-1} a_{ij} x_j^{k+1} = (1 - \omega)a_{ii} x_i^k - \omega \sum_{j=i+1}^{n} a_{ij} x_j^k + \omega b_i. \tag{13}$$

This relation of the new iterates x_i^{k+1} to the old x_i^k holds for $i = 1, \ldots, n$ and, by means of the decomposition (7), may be written as

$$D\mathbf{x}^{k+1} - \omega L\mathbf{x}^{k+1} = (1 - \omega)D\mathbf{x}^k + \omega U\mathbf{x}^k + \omega\mathbf{b} \tag{14}$$

or, under the assumption (2),

$$\mathbf{x}^{k+1} = H_\omega \mathbf{x}^k + \omega(D - \omega L)^{-1}\mathbf{b}, \qquad k = 0, 1, \ldots \tag{15}$$

where we have set

$$H_\omega = (D - \omega L)^{-1}[(1 - \omega)D + \omega U]. \tag{16}$$

Again, it is evident that (15) reduces to the Gauss–Seidel iteration (8) when $\omega = 1$.

We first prove a basic result which gives the maximum range of values of ω for which the SOR iteration can converge.

7.1.6 (Kahan's Theorem)   Assume that $A \in L(C^n)$ has nonzero diagonal elements. Then

$$\rho(H_\omega) \geq |\omega - 1|. \tag{17}$$

Proof: Because L is strictly lower triangular, $\det D^{-1} = \det(D - \omega L)^{-1}$ and we have

$$\det H_\omega = \det(D - \omega L)^{-1} \det\{(1 - \omega)D + \omega U\}$$
$$= \det\{(1 - \omega)I + \omega D^{-1}U\} = \det\{(1 - \omega)I\} = (1 - \omega)^n$$

since $D^{-1}U$ is strictly upper triangular. But $\det H_\omega$ is the product of the eigenvalues of H_ω and therefore (17) must hold. $\$\$\$

In order that the iterates (15) converge for all $\mathbf{x}^0$, it is necessary, by 7.1.1, that $\rho(H_\omega) < 1$ and hence, by (17), that $0 < \omega < 2$ if ω is real. For an important class of matrices, we shall show that this is also a sufficient condition. We first introduce another class of splittings.

7.1.7 Definition†   Let A, B, and C be in $L(R^n)$. Then $A = B - C$ is a **P-regular** splitting of A if B is nonsingular and $B + C$ is positive definite.

Note that this definition does not require that either A or $P = B + C$ be symmetric. Rather, we want that $\mathbf{x}^T P \mathbf{x} > 0$ for all nonzero $\mathbf{x} \in R^n$, which is equivalent to the requirement that the **symmetric part** of P, defined by $\frac{1}{2}(P + P^T)$, be positive definite (E7.1.5).

The following result is the basis for the convergence theorems to follow.

7.1.8 (Stein's Theorem)   Let $H \in L(R^n)$ and assume that $A \in L(R^n)$ is a symmetric positive definite matrix such that $A - H^T A H$ is positive definite. Then $\rho(H) < 1$.

Proof: Let λ be any eigenvalue of H and $\mathbf{u} \neq \mathbf{0}$ a corresponding eigenvector. Then $\mathbf{u}^H A \mathbf{u}$ and $\mathbf{u}^H(A - H^T A H)\mathbf{u}$ are real (E7.1.6) and positive and therefore

$$\mathbf{u}^H A \mathbf{u} > \mathbf{u}^H H^T A H \mathbf{u} = (\lambda \mathbf{u})^H A(\lambda \mathbf{u}) = |\lambda|^2 \mathbf{u}^H A \mathbf{u},$$

so that $|\lambda|^2 < 1$. $\$\$\$

† This definition does not appear to be in the literature. It is motivated by Theorem 7.1.9 which was apparently first proved J. Weissinger, *Z. Angew. Math. Mech.* **33** (1953), 155–163.

7.1.9   (P-Regular Splitting Theorem)   Let $A \in L(R^n)$ be symmetric positive definite and $A = B - C$ a P-regular splitting. Then $\rho(B^{-1}C) < 1$.

Proof: By 7.1.8, it suffices to show that

$$Q = A - (B^{-1}C)^{T}AB^{-1}C$$

is positive definite. Since $B^{-1}C = I - B^{-1}A$, we have

$$Q = (B^{-1}A)^{T}A + AB^{-1}A - (B^{-1}A)^{T}AB^{-1}A$$
$$= (B^{-1}A)^{T}(B + B^{T} - A)B^{-1}A.$$

But $B + B^{T} - A = B^{T} + C$ is positive definite with $B + C$ and therefore Q is positive definite (E7.1.7). $\$\$\$$

For the SOR iteration 7.1.9 reduces to the following famous result.

7.1.10   (Ostrowski–Reich Theorem)   Let $A \in L(R^n)$ be symmetric positive definite and assume that $0 < \omega < 2$. Then the SOR iterates (15) converge to $A^{-1}\mathbf{b}$ for any $\mathbf{x}^0$ and $\mathbf{b} \in R^n$.

Proof: By 7.1.1 and 7.1.9, it suffices to show that

$$A = \omega^{-1}(D - \omega L) - \omega^{-1}[(1 - \omega)D + \omega L^{T}]$$

is a P-regular splitting of A. Since the diagonal elements of A are positive, D is positive definite and $D - \omega L$ is nonsingular. Moreover, the symmetric part of $B + C$ is

$$B + B^{T} - A = 2\omega^{-1}D - L - L^{T} - D + L + L^{T} = \omega^{-1}(2 - \omega)D$$

which, since $0 < \omega < 2$, is positive definite. $\$\$\$$

We note that 7.1.8–7.1.10 all have interesting converses. For these we refer to E7.1.8 and E7.1.9.

EXERCISES

E7.1.1   Consider the iterative process (9) and assume that the equation $\mathbf{x} = H\mathbf{x} + \mathbf{d}$ has a unique solution $\mathbf{x}^*$. Show that the sequence $\{\mathbf{x}^k\}$ converges to $\mathbf{x}^*$ if and only if $\mathbf{x}^0 - \mathbf{x}^*$ lies in a subspace spanned by eigenvectors and principal vectors of H associated with eigenvalues of modulus less than unity.

E7.1.2　Show that the following converse of 7.1.3 holds: If $A = B - C$ is a regular splitting of A and $\rho(B^{-1}C) < 1$, then A^{-1} exists and is nonnegative.

E7.1.3　$A = B - C$ is a **weak regular splitting** of $A \in L(R^n)$ if B^{-1} exists and is nonnegative, and $B^{-1}C \geq 0$ and $CB^{-1} \geq 0$. Show that 7.1.3 and also the converse in E7.1.2 remain valid for weak regular splittings.

E7.1.4　Let $A \in L(R^n)$ be an M-matrix. Modify the proof of 7.1.4 to show that $\rho(H_\omega) < 1$ for $\omega \in (0, 1]$, where H_ω is given by (16).

E7.1.5　Let $A \in L(R^n)$. Show that $\mathbf{x}^T A \mathbf{x} \geq 0$ for all $\mathbf{x} \in R^n$ if and only if $\mathbf{x}^T (A + A^T)\mathbf{x} \geq 0$ for all $\mathbf{x} \in R^n$.

E7.1.6　Let $A \in L(R^n)$. Show that $A = A_s + A_a$ where A_s is symmetric and A_a is skew-symmetric. Use this to show that $\mathbf{u}^H A \mathbf{u}$ is real for every complex vector $\mathbf{u}$ if and only if A is symmetric.

E7.1.7　Let $A \in L(R^n)$ be symmetric positive definite and $H \in L(R^n)$ nonsingular. Show that $H^T A H$ is positive definite.

E7.1.8　Prove the following converse of 7.1.8: If $\rho(H) < 1$, there exists an hermitian positive definite $A \in L(C^n)$ such that $A - H^H A H$ is positive definite. (*Hint:* Choose P such that $\|PHP^{-1}\|_2 < 1$ and set $A = P^H P$. Then show that $\|PH\mathbf{x}\|_2 < \|P\mathbf{x}\|_2$ for all $\mathbf{x} \in C^n$.)

E7.1.9　Prove the following converse of 7.1.9. If $A \in L(R^n)$ is symmetric and $A = B - C$ is a P-regular splitting with $\rho(B^{-1}C) < 1$, then A is positive definite. Similarly, show that if $A \in L(R^n)$ is symmetric with positive diagonal elements and $\rho(H_\omega) < 1$ for some $\omega \in (0, 2)$, then A is positive definite. [*Hint:* If $\mathbf{x}_{k+1} = B^{-1}C\mathbf{x}_k$, $k = 0, 1, \ldots$, show that

$$\mathbf{x}_k^T A \mathbf{x}_k - \mathbf{x}_{k+1}^T A \mathbf{x}_{k+1} = (\mathbf{x}_k - \mathbf{x}_{k+1})^T (B + C)(\mathbf{x}_k - \mathbf{x}_{k+1})$$

for all $k \geq 1$, and hence if $\mathbf{x}_0^T A \mathbf{x}_0 \leq 0$ for some $\mathbf{x}_0$ the convergence of $\{\mathbf{x}_k\}$ to zero would be contradicted.]

E7.1.10　Show that the matrix

$$A = \begin{bmatrix} 1 & a & a \\ a & 1 & a \\ a & a & 1 \end{bmatrix}$$

is positive definite for $-\frac{1}{2} < a < 1$ but that the Jacobi iterates (4) converge only for $-\frac{1}{2} < a < \frac{1}{2}$.

7.2 RATE OF CONVERGENCE

The introduction of the parameter ω into the Gauss–Seidel iteration, discussed in the previous section, is not done so as to force convergence but, rather, to enhance the rate of convergence. In this section, we will investigate primarily the optimum choice of ω in this regard. We first need to make precise the concept of rate of convergence, and we do this for an arbitrary iteration of the form

$$\mathbf{x}^{k+1} = G\mathbf{x}^k, \qquad k = 0, 1, \ldots \tag{1}$$

where G is a (perhaps nonlinear) mapping from R^n to R^n.

7.2.1 Definition Assume that $\mathbf{x}^* = G\mathbf{x}^*$ and let C be the set of all sequences $\{\mathbf{x}^k\}$ generated by (1) such that $\lim_{k \to \infty} \mathbf{x}^k = \mathbf{x}^*$. Then

$$\alpha = \sup\left\{\limsup_{k \to \infty} \|\mathbf{x}^k - \mathbf{x}^*\|^{1/k} : \{\mathbf{x}^k\} \in C\right\} \tag{2}$$

is the **asymptotic convergence factor** of the iteration (1) at $\mathbf{x}^*$.

In order to understand better the role of α, consider a single sequence $\{\mathbf{x}^k\}$ which converges to $\mathbf{x}^*$ and let

$$\beta = \limsup_{k \to \infty} \|\mathbf{x}^k - \mathbf{x}^*\|^{1/k}. \tag{3}$$

Since $\lim_{k \to \infty}(\mathbf{x}^k - \mathbf{x}^*) = 0$, clearly $0 \le \beta \le 1$, and thus the same is true of α. From (3), it follows that for any $\varepsilon > 0$ there is a k_0 such that

$$\|\mathbf{x}^k - \mathbf{x}^*\| \le (\beta + \varepsilon)^k \tag{4}$$

for all $k \ge k_0$. Hence, if $\beta < 1$, we may choose ε so that $\beta + \varepsilon < 1$ and (4) shows that, asymptotically, $\|\mathbf{x}^k - \mathbf{x}^*\|$ tends to zero at least as rapidly as the geometric sequence $(\beta + \varepsilon)^k$. The supremum in (2) is taken so as to reflect the worst possible rate of convergence of any individual sequence. Clearly, the smaller α is, the faster the rate of convergence of the process. We also note that α is independent of the particular norm used (E7.2.1).

For a linear iterative process

$$\mathbf{x}^{k+1} = H\mathbf{x}^k + \mathbf{d}, \qquad k = 0, 1, \ldots \tag{5}$$

it is possible to give a simple characterization of α.

7.2.2 Let $H \in L(R^n)$ and assume that $\rho(H) < 1$. Then the asymptotic convergence factor of the iteration (5) is $\rho(H)$.

Proof: For given $\varepsilon > 0$, 1.3.6 ensures that there is a norm such that $\|H\| \leq \rho(H) + \varepsilon$. Hence, if $x^* = Hx^* + d$, then

$$\|x^k - x^*\| \leqslant \|H\|^k \|x^0 - x^*\| \leqslant [\rho(H) + \varepsilon]^k \|x^0 - x^*\|, \qquad k = 0, 1, \ldots.$$

Since $a^{1/k} \to 1$ as $k \to \infty$ for any $a > 0$, we have

$$\limsup_{k \to \infty} \|x^k - x^*\|^{1/k} \leq \rho(H) + \varepsilon$$

so that $\alpha \leq \rho(H) + \varepsilon$. But since ε is arbitrary and α is norm independent (E7.2.1), it follows that $\alpha \leq \rho(H)$. To show that equality holds it suffices to exhibit a single sequence for which

$$\limsup_{k \to \infty} \|x^k - x^*\|^{1/k} = \rho(H). \tag{6}$$

Suppose that there is a real eigenvalue λ such that $|\lambda| = \rho(H)$. If we choose x^0 such that $x^0 - x^*$ is in the direction of an eigenvector of H associated with λ, then

$$\|x^k - x^*\| = \|H^k(x^0 - x^*)\| = \|\lambda^k(x^0 - x^*)\| = \rho(H)^k \|x^0 - x^*\|$$

so that (6) holds. On the other hand,† if there are no real eigenvalues of modulus $\rho(H)$, then, since H is real, there is a complex conjugate pair λ and $\bar{\lambda}$ of eigenvalues such that $|\lambda| = \rho(H)$ with (necessarily complex) eigenvectors u and $\bar{u}$. We may extend $u_1 = u$ and $u_2 = \bar{u}$ to a basis $u_1, \ldots, u_n$ for the complex space C^n and, hence, any vector $y \in R^n$ may be written as $y = \sum_{i=1}^n \hat{y}_i u_i$ where the coefficients $\hat{y}_i$ may be complex. Then $\|y\| = \sum_{i=1}^n |\hat{y}_i|$ defines a norm on R^n (E7.2.2). Now choose x^0 so that $x^0 - x^* = \text{Re } u = \frac{1}{2}(u + \bar{u})$, which is necessarily nonzero (E7.2.3). Then

$$x^k - x^* = H^k(x^0 - x^*) = \tfrac{1}{2}H^k(u + \bar{u}) = \tfrac{1}{2}(\lambda^k u_1 + \bar{\lambda}^k u_2).$$

Hence, in the norm previously defined,

$$\|x^k - x^*\| = \tfrac{1}{2}(|\lambda^k| + |\bar{\lambda}^k|) = \rho(H)^k, \qquad k = 0, 1, \ldots,$$

so that, clearly, (6) holds. $\$\$\$$

We return now to the SOR iteration (7.1.15) and pose the problem of choosing the parameter ω so as to maximize the rate of convergence. More

† The rest of this proof would not be necessary if we could choose $x^0 \in C^n$.

precisely, we shall seek to minimize the convergence factor α of 7.2.1 and, by 7.2.2, this is equivalent to minimizing the spectral radius of

$$H_\omega = (D - \omega L)^{-1}[(1 - \omega)D + \omega U], \tag{7}$$

where, as usual, D, $-L$, and $-U$ are the diagonal, strictly lower triangular, and strictly upper triangular parts of the coefficient matrix A. We will be able to give a complete solution to this problem for a certain class of matrices. Throughout this section we will assume, of course, that A has nonzero diagonal elements so that H_ω is well defined.

The Gauss–Seidel method, and hence also the SOR method, is dependent upon the ordering of the equations. Of these $n!$ possible orderings, we will consider only those for which the corresponding coefficient matrix is of the following type.

7.2.3 Definition The matrix $A \in L(R^n)$ is **consistently ordered**† if the eigenvalues of

$$B(\alpha) \equiv \alpha^{-1}D^{-1}L + \alpha D^{-1}U$$

are independent of α for all $\alpha \neq 0$.

As an example of a consistently ordered matrix consider the matrix

$$A = \begin{bmatrix} 2 & -1 & & \bigcirc \\ -1 & 2 & \ddots & \\ & \ddots & \ddots & -1 \\ \bigcirc & & -1 & 2 \end{bmatrix} \tag{8}$$

which arose in the previous chapter. Here

$$B(\alpha) = \frac{1}{2\alpha} \begin{bmatrix} 0 & \alpha^2 & & \bigcirc \\ 1 & \ddots & \ddots & \\ & \ddots & \ddots & \alpha^2 \\ \bigcirc & & 1 & 0 \end{bmatrix}$$

† See N. Nichols and L. Fox, *Numer. Math.* **13** (1969), 425–433, and Young [1971] for other definitions.

and it is easy to see (E6.2.5) that

$$SB(\alpha)S^{-1} = \frac{1}{2}\begin{bmatrix} 0 & 1 & & & \\ 1 & \cdot & \cdot & & \text{\Large O} \\ & \cdot & \cdot & \cdot & \\ & & \cdot & \cdot & 1 \\ \text{\Large O} & & & 1 & 0 \end{bmatrix}$$

where $S = \text{diag}(1, \alpha, \ldots, \alpha^{n-1})$. Hence $B(\alpha)$ is similar to a matrix which is independent of α and therefore the eigenvalues of $B(\alpha)$ itself are independent of α. Thus A is consistently ordered.

Along with consistent ordering, we shall need another property.

7.2.4 Definition The matrix $A \in L(R^n)$ is **2-cyclic** (or has **property A**) if there is a permutation matrix P such that

$$PAP^T = \begin{bmatrix} D_1 & C_1 \\ C_2 & D_2 \end{bmatrix}, \tag{9}$$

where D_1 and D_2 are diagonal.

Consider again the matrix (8) and recall that it arose in Chapter 6 from the two-point boundary value problem $y'' = 0$, $y(a) = \alpha$, $y(b) = \beta$ with grid points $a + ih$, $i = 0, \ldots, n + 1$. Now label the grid points alternately red and black and renumber the red points $0, 1, \ldots, q$ and the black points $q + 1, \ldots, n + 1$, as shown in Figure 7.2.1. Then the difference equations

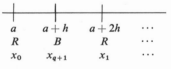

Figure 7.2.1

are of the form, for even n †

$$x_i - 2x_{q+i+1} + x_{i+1} = 0, \qquad i = 0, \ldots, q - 1$$

$$x_{q+i} - 2x_i + x_{q+i+1} = 0, \qquad i = 1, \ldots, q$$

† For odd n, an obvious modification is necessary.

or

$$
\begin{bmatrix}
2 & & & & -1 & -1 & & \\
& \ddots & & & & \ddots & \ddots & \\
& & & & & & \ddots & -1 \\
& & 2 & & & & & -1 \\
\hline
-1 & & & 2 & & & & \\
-1 & \ddots & & & \ddots & & & \\
& \ddots & \ddots & & & & & \\
& & -1 & -1 & & & \cdot & 2
\end{bmatrix}
\begin{bmatrix}
x_1 \\
\vdots \\
\\
x_q \\
\hline
x_{q+1} \\
\\
\vdots \\
x_n
\end{bmatrix}
=
\begin{bmatrix}
0 \\
\vdots \\
0 \\
x_{n+1} \\
\hline
x_0 \\
0 \\
\vdots \\
0
\end{bmatrix}
$$

The renumbering of the grid points corresponds to a transformation PAP^{T} of A where P is a permutation matrix. Hence A is 2-cyclic.

We can now begin our analysis. We first prove a basic lemma which shows that the eigenvalues of the Jacobi iteration matrix $J = D^{-1}(L + U)$ occur in pairs, $\pm\mu_i$, with the possible exception of the zero eigenvalue.

7.2.5 (Romanovsky's Lemma)   Assume that $A \in L(R^n)$ is 2-cyclic with nonzero diagonal elements. Then there are nonnegative integers p and r with $p + 2r = n$ such that

$$\det(\lambda I - J) = \lambda^p \prod_{i=1}^{r} (\lambda^2 - \mu_i^2) \tag{10}$$

for certain (real or complex) numbers $\mu_1, \ldots, \mu_r$.

Proof: It suffices to prove that if μ is an eigenvalue of J then $-\mu$ is also. This, in turn, is proved if $\det(\lambda I - J) = \det(\lambda I + J)$ for any λ. Now since A is 2-cyclic, there is a permutation matrix P so that (9) holds. Hence

$\det(\lambda I - J) = \det P \det(\lambda I - J) \det P^{\mathrm{T}}$

$$
\begin{aligned}
&= \det \begin{bmatrix} \lambda I_1 & D_1^{-1} C_1 \\ D_2^{-1} C_2 & \lambda I_2 \end{bmatrix} \\
&= (-1)^n \det \begin{bmatrix} -I_1 & \\ & I_2 \end{bmatrix} \det \begin{bmatrix} \lambda I_1 & D_1^{-1} C_1 \\ D_2^{-1} C_2 & \lambda I_2 \end{bmatrix} \det \begin{bmatrix} I_1 & \\ & -I_2 \end{bmatrix} \\
&= (-1)^n \det \begin{bmatrix} -\lambda I_1 & D_1^{-1} C_1 \\ D_2^{-1} C_2 & -\lambda I_2 \end{bmatrix} = \det(\lambda I + J), \tag{11}
\end{aligned}
$$

and the result is proved. $\$\$\$$

We note that if D_i is $r_i \times r_i$ it may be shown that the integer r in (10) is just the smaller of the r_i.

We next need another lemma which utilizes the concept of consistent ordering.

7.2.6 Let $A \in L(R^n)$ be consistently ordered with nonzero diagonal elements. Then for any constants† α, β, and γ,

$$\det\{\gamma D - \alpha L - \beta U\} = \det\{\gamma D - \alpha^{1/2}\beta^{1/2}(L + U)\}. \tag{12}$$

Proof: If we multiply (12) by $\det D^{-1}$, then it is clear that the result is proved if we show that the eigenvalues of $\alpha D^{-1}L + \beta D^{-1}U$ are identical with those of $\alpha^{1/2}\beta^{1/2}D^{-1}(L + U)$. If either α or β is zero, then the first matrix is strictly triangular while the second is zero so that the eigenvalues of both are all zero. Hence we may assume that neither α nor β is zero. Set $\delta = \alpha^{1/2}/\beta^{1/2}$. Then

$$\alpha D^{-1}L + \beta D^{-1}U = \alpha^{1/2}\beta^{1/2}\{\delta D^{-1}L + \delta^{-1}D^{-1}U\}.$$

By the hypothesis of consistent ordering, the eigenvalues of the matrix on the right are independent of δ; hence we may set $\delta = 1$ and conclude that the eigenvalues of $\alpha D^{-1}L + \beta D^{-1}U$ are the same as those of $\alpha^{1/2}\beta^{1/2}\{D^{-1}L + D^{-1}U\}$. $\$\$\$$

We are now able to prove a basic result which relates the eigenvalues of the SOR iteration matrix (7) to those of the Jacobi iteration matrix $J = D^{-1}(L + U)$.

7.2.7 Assume that $A \in L(R^n)$ is consistently ordered and 2-cyclic with nonzero diagonal elements, and that $\omega \geq 0$. Then there are nonnegative integers p and r with $p + 2r = n$ and (real or complex) numbers $\mu_1, \ldots, \mu_r$ such that

$$\det(\mu I - J) = \mu^p \prod_{i=1}^{r} (\mu^2 - \mu_i^2) \tag{13}$$

and

$$\det(\lambda I - H_\omega) = (\lambda + \omega - 1)^p \prod_{i=1}^{r} [(\lambda + \omega - 1)^2 - \lambda\omega^2\mu_i^2]. \tag{14}$$

† Real or complex; it is not necessary that $\alpha, \beta \geq 0$.

Proof: Since $D - \omega L$ is nonsingular, and det $D^{-1} = \det(D - \omega L)^{-1}$, we have, using **7.2.6**,

$$\det(\lambda I - H_\omega) = \det(D - \omega L)^{-1} \det\{(D - \omega L)\lambda - (1 - \omega)D - \omega U\}$$
$$= \det\{(\lambda + \omega - 1)I - \lambda \omega D^{-1}L - \omega D^{-1}U\}$$
$$= \det\{(\lambda + \omega - 1)I - \lambda^{1/2}\omega J\}.$$

The representation (13) is a direct consequence of **7.2.5** as is (14) since the eigenvalues of $\lambda^{1/2}\omega J$ are scalar multiples of those of J and the variable $\lambda + \omega - 1$ now plays the role of λ. $\$\$\$$

The importance of the last result is that it shows a direct correspondence between the eigenvalues of H_ω and those of J. In particular, if J has a p-fold zero eigenvalue, then H_ω has p corresponding eigenvalues equal to $1 - \omega$. Moreover, associated with the $2r$ nonzero eigenvalues $\pm \mu_i$ of J are $2r$ eigenvalues of H_ω which satisfy

$$(\lambda_i + \omega - 1)^2 = \lambda_i \omega^2 \mu_i^2. \tag{15}$$

We summarize these relationships in Table 1.

TABLE 1   Eigenvalues of J and H_ω

J	H_1	H_ω
0	0	$1 - \omega$
$\pm \mu_i$	$0, \mu_i^2$	given by (15)

As indicated in Table 1, an immediate consequence of **7.2.7** is:

7.2.8   Under the conditions of **7.2.7**, $\rho(H_1) = \rho(J)^2$.

If $\rho(J) < 1$, so that the Jacobi iteration is convergent, then **7.2.8** shows that the Gauss–Seidel iteration is also convergent and, more importantly, that the asymptotic convergence factor of the Gauss–Seidel iteration is the square of that of the Jacobi iteration; that is, in terms of the asymptotic convergence factors the Gauss–Seidel iteration is precisely twice as fast as the Jacobi iteration.

We will next use **7.2.7** to choose ω so as to minimize $\rho(H_\omega)$ and hence maximize the rate of convergence of the SOR iteration. *In the sequel, we shall assume, in addition to the conditions of **7.2.7**, that the eigenvalues of J are real and $\rho(J) < 1$.* This is the case, for example, for the matrix (8).

We write (15) in the form

$$(\lambda_i + \omega - 1)/\omega = \pm \lambda^{1/2} \mu_i \tag{16}$$

and plot each side of this equation separately as a function of λ_i for fixed values of $\omega \in (0, 2)$ and μ_i. This is shown in Figure 7.2.2 where we have

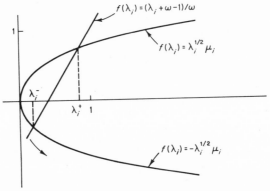

Figure 7.2.2

denoted the roots of (16) as λ_i- and λ_i+ corresponding to the values $\pm \mu_i$. As ω decreases from 1 to 0, it is clear that the line $(\lambda_i + \omega - 1)/\omega$, with slope ω^{-1}, rotates about $(1, 1)$ toward a vertical position and, consequently the roots λ_i+ and λ_i- both increase monotonically. This accounts for the behavior of all the eigenvalues of H_ω which satisfy (16). The other (possible) eigenvalues $\lambda = 1 - \omega$ also increase monotonically as ω decreases from 1 to 0 and hence the same is true of $\rho(H_\omega)$; that is,

$$\rho(H_{\omega_1}) < \rho(H_{\omega_2}) \qquad \text{if} \quad \omega_2 < \omega_1 \leq 1. \tag{17}$$

Therefore, the optimum ω in the interval $(0, 1]$ is $\omega = 1$.

As ω increases from 1, the line $(\lambda_i + \omega - 1)/\omega$ now swings upward (see Figure 7.2.3) and the root λ_i- increases while the root λ_i+ decreases until

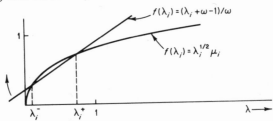

Figure 7.2.3

they coalesce at the point that $(\lambda_i + \omega - 1)/\omega$ is tangent to the curve $f(\lambda) = \lambda_i^{1/2}\mu_i$. At this point of tangency, the slopes of the two curves as well as the function values are equal; this gives

$$2\lambda_i^{1/2} = \mu_i\omega, \qquad \lambda_i + \omega - 1 = \lambda_i^{1/2}\mu_i\omega. \qquad (18)$$

If we eliminate λ_i in these relations, we see that ω must satisfy the quadratic equation

$$\mu_i^2\omega^2 - 4\omega + 4 = 0$$

so that

$$\omega = \frac{2 - 2\sqrt{1 - \mu_i^2}}{\mu_i^2} = \frac{2}{1 + \sqrt{1 - \mu_i^2}}. \qquad (19)$$

As ω increases beyond this value, it is clear that the root pair $\lambda_i\pm$ becomes complex. In fact, if we regard (16) as a quadratic in $\lambda_i^{1/2}$, then [with the plus sign in (16)]

$$\lambda_i^{1/2} = [\omega\mu_i \pm \sqrt{\omega^2\mu_i^2 - 4(\omega - 1)}]/2$$

where the discriminant $\omega^2\mu_i^2 - 4\omega + 4$ is negative if ω exceeds the value of (19). Hence

$$|\lambda_i| = \lambda_i^{1/2}\bar{\lambda}_i^{1/2} = [\omega^2\mu_i^2 - (\omega^2\mu_i^2 - 4\omega + 4)]/4 = \omega - 1. \qquad (20)$$

If we now consider the root pairs $\lambda_i\pm$ corresponding to $\mu_1^2 \leq \mu_2^2 \leq \cdots \leq \mu_r^2$ as ω increases from 1, we see that $\lambda_1\pm$ first becomes complex with modulus $\omega - 1$, then $\lambda_2\pm$ becomes complex, and so forth until finally $\lambda_r\pm$ becomes complex at the value of ω given by (19):

$$\omega_0 = \frac{2}{1 + \sqrt{1 - \rho(J)^2}}. \qquad (21)$$

The p-fold root $\omega - 1$ of course has the absolute value $\omega - 1$ also so at the value of ω_0 given by (21) all of the eigenvalues of H_ω have modulus $\omega_0 - 1$ and as ω increases further all eigenvalues have modulus $\omega - 1$, as shown by (20). For values of $\omega < \omega_0$, λ_r+ is still decreasing as ω increases. Hence $\rho(H_\omega)$ is minimized at the value of ω given by (21). The behavior of the roots of H_ω in the complex plane is shown in Figure 7.2.4.

We summarize the above discussion in the following result. Note that a sufficient condition that the roots of J be real is that A be symmetric and positive definite (E7.2.7) and that sufficient conditions for $\rho(J) < 1$ were given in the previous section (7.1.4, 7.1.5).

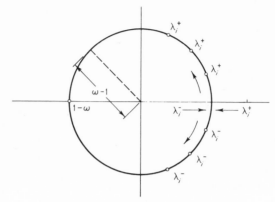

Figure 7.2.4   Roots of H_ω in complex plane.

7.2.9   Assume that $A \in L(R^n)$ is consistently ordered and 2-cyclic with nonzero diagonal elements. Assume further that the eigenvalues of the Jacobi iteration matrix J are real and that $\rho(J) < 1$. Then for any $\omega \in (0, 2)$, $\rho(H_\omega) < 1$ and there is a unique value of ω, given by (21), for which

$$\rho(H_{\omega_0}) = \min_{0 < \omega < 2} \rho(H_\omega).$$

In Figure 7.2.5, we plot $\rho(H_\omega)$ as a function of ω. It can be shown that as ω approaches ω_0 from the left the slope of the curve becomes infinite,

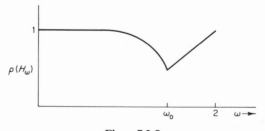

Figure 7.2.5

while to the right of ω_0 the slope is 1. Hence, it is a better computational tactic to approximate ω_0 by a value which is too large rather than too small.

EXERCISES

E7.2.1 Use the norm equivalence theorem 1.2.4 to show that if $\mathbf{x}^k \to \mathbf{x}^*$ as $k \to \infty$, then

$$\limsup_{k \to \infty} \|\mathbf{x}^k - \mathbf{x}^*\|^{1/k}$$

is independent of the norm. Hence conclude that the convergence factor α of 7.2.1 is independent of the norm.

E7.2.2 Let $\mathbf{u}_1, \ldots, \mathbf{u}_n$ be a basis for C^n and for any $\mathbf{y} \in R^n$ let $\mathbf{y} = \sum_{i=1}^n \hat{y}_i \mathbf{u}_i$. Show that $\|\mathbf{y}\| = \sum_{i=1}^n |\hat{y}_i|$ is a norm on R^n.

E7.2.3 Let $H \in L(R^n)$. Show that if $\mathbf{u}$ is a (nonzero) eigenvector of H corresponding to the complex eigenvalue λ, then Re $\mathbf{u} \neq \mathbf{0}$.

E7.2.4 Consider the matrices

$$A_1 = \begin{bmatrix} \frac{1}{2} & 0 \\ 0 & \frac{1}{2} \end{bmatrix}, \qquad A_2 = \begin{bmatrix} \frac{1}{4} & \beta \\ 0 & \frac{1}{4} \end{bmatrix}.$$

Show that $\alpha(A_2) = \frac{1}{4} < \alpha(A_1) = \frac{1}{2}$, where α is the convergence factor of 7.2.1, but that given any number N, we can choose β and $\mathbf{x}^0$ so that

$$\|A_1^k \mathbf{x}^0\| < \|A_2^k \mathbf{x}^0\| \qquad \text{for} \quad k = 0, \ldots, N.$$

E7.2.5 Use E6.1.1 to compute the optimum ω for the matrix (8).

E7.2.6 Show that the matrix of (9) is consistently ordered.

E7.2.7 Assume that B, $C \in L(R^n)$ are symmetric and that B is positive definite. Show that the eigenvalues of BC are real. Use this to show that if $A \in L(R^n)$ is symmetric positive definite then the eigenvalues of its Jacobi iteration matrix are real.

7.3 APPLICATIONS TO DIFFERENTIAL EQUATIONS

We will now apply the results of the previous two sections to discrete analogues of differential equations. Consider first the system of equations

$$-y_{i-1} + 2y_i - y_{i+1} + h^2 q_i y_i + h^2 r_i = 0, \qquad i = 1, \ldots, n \qquad (1)$$

where $y_0 = \alpha$, $y_{n+1} = \beta$, h, q_1, ..., q_n, and r_1, ..., r_n are known. As discussed in Chapter 6, this system arises from the two-point boundary value problem

$$y''(x) = q(x)y(x) + r(x), \qquad y(a) = \alpha, \quad y(b) = \beta.$$

As in Chapter 6, we write (1) in the form

$$A\mathbf{x} = \mathbf{b} \tag{2}$$

where

$$A = \begin{bmatrix} 2 + h^2 q_1 & -1 & & & \\ -1 & 2 + h^2 q_2 & \cdot & & \\ & \cdot & \cdot & \cdot & \\ & & \cdot & \cdot & -1 \\ & & & -1 & 2 + h^2 q_n \end{bmatrix}, \quad \mathbf{b} = \begin{bmatrix} -h^2 r_1 + \alpha \\ -h^2 r_2 \\ \vdots \\ -h^2 r_{n-1} \\ -h^2 r_n + \beta \end{bmatrix} \tag{3}$$

In Theorem 6.2.8, we showed that A was an M-matrix under the condition that

$$q_i \geq 0, \qquad i = 1, \ldots, n. \tag{4}$$

Hence, Theorem 7.1.4 applies and ensures that either the Jacobi iterates (7.1.4) or the Gauss–Seidel iterates (7.1.8) converge to the unique solution of (2) for any starting approximation $\mathbf{x}^0$. Moreover, since A is symmetric, 6.2.15 shows that A is positive definite. Hence, by the Ostrowski–Reich Theorem 7.1.10 we may conclude that the SOR iterates (7.1.15) also converge to $A^{-1}\mathbf{b}$ for any $\omega \in (0, 2)$. Finally, the same analysis as used in 7.2 can be used to show that A is 2-cyclic and consistently ordered (E7.3.1). Moreover, it is also easy to see (E7.3.2) that the eigenvalues of the Jacobi iteration matrix for A are real. Hence Theorem 7.2.8 holds and an optimum ω exists for the SOR iteration applied to (2).

Although, as we have just seen, all of the theoretical results of Sections 7.1 and 7.2 apply to the equation (2), it is far better in practice to use gaussian elimination (see Chapter 9) to solve this type of linear system. For partial differential equations, however, the situation is somewhat different and iterative methods are, in most cases, to be preferred over direct methods.†

---

† Although direct methods seem to be increasingly competitive. See, for example, B. Buzbee, G. Golub, and C. Neilson, On Direct Methods for Solving Poisson's Equations, *SIAM J. on Numer. Anal.* **7** (1970), 627–656.

We shall now illustrate how the results of the two previous sections apply to what is probably the simplest situation: Laplace's equation on a square. We consider the equation

$$\Delta u(s, t) \equiv u_{ss}(s, t) + u_{tt}(s, t) = 0, \qquad s, t \in \Omega, \tag{5}$$

where Ω is the unit square $[0, 1] \times [0, 1]$, and u is to satisfy on the boundary $\dot{\Omega}$ of Ω the condition

$$u(s, t) = g(s, t), \qquad s, t \in \dot{\Omega} \tag{6}$$

where g is a given function. We can obtain a discrete analogue for (5) in a manner analogous to that used in Chapter 6 for ordinary differential equations. First, we divide Ω by a uniform square mesh with spacing h and define the **grid points** $P_{ij} = (ih, jh)$, $i, j = 0, \ldots, m + 1$, where $(m + 1)h = 1$; this is depicted in Figure 7.3.1 for $h = \frac{1}{3}$. At each interior

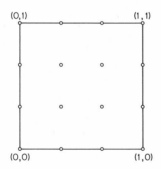

(0,1)   (1,1)

(0,0)   (1,0)

Figure 7.3.1

grid point, the second partial derivatives u_{ss} and u_{tt} are now approximated by central difference quotients; that is,

$$u_{ss}(ih, jh) \doteq [U_{i+1, j} - 2U_{ij} + U_{i-1, j}]/h^2$$

$$u_{tt}(ih, jh) \doteq [U_{i, j+1} - 2U_{ij} + U_{i, j-1}]/h^2$$

where we have set $U_{ij} = u(ih, jh)$. If we use these approximations in (5), we then obtain the discrete analogue

$$(U_{i+1, j} - 2U_{ij} + U_{i-1, j})/h^2 + (U_{i, j+1} - 2U_{ij} + U_{i, j-1})/h^2 = 0$$

or, upon multiplying through by $-h^2$,

$$4U_{ij} - U_{i-1, j} - U_{i+1, j} - U_{i, j+1} - U_{i, j-1} = 0, \qquad j = 1, \ldots, m. \tag{7}$$

Since the values of U_{ij} at the boundary grid points are assumed to be known by (6), this is a system of $n = m^2$ equations in the unknowns U_{ij}, $i, j = 1, \ldots, m$.

In order to write (7) in matrix form, we associate a vector $\mathbf{x} \in R^n$ with the unknowns U_{ij} by the correspondence

$$x_1 = U_{11}, \ldots, x_m = U_{m1}, \quad x_{m+1} = U_{12}, \ldots, x_n = U_{mm}$$

and define the block tridiagonal matrix

$$A = \begin{bmatrix} B & -I & & & \bigcirc \\ -I & \cdot & \cdot & & \\ & \cdot & \cdot & \cdot & \\ & & \cdot & \cdot & -I \\ \bigcirc & & & -I & B \end{bmatrix}, \tag{8}$$

where I is the $m \times m$ identity matrix and B is the $m \times m$ matrix

$$B = \begin{bmatrix} 4 & -1 & & & \bigcirc \\ -1 & \cdot & \cdot & & \\ & \cdot & \cdot & \cdot & \\ & & \cdot & \cdot & -1 \\ \bigcirc & & & -1 & 4 \end{bmatrix}. \tag{9}$$

Then (7) may be written as

$$A\mathbf{x} = \mathbf{b} \tag{10}$$

where the components of the vector $\mathbf{b}$ are either zero or are the known boundary values (E7.3.3).

We next show that the matrix A is irreducible and for this we will apply Theorem **6.2.2**, which states that A is irreducible if for any indices $1 \le i < j \le n$ there is a sequence of nonzero elements of A of the form $a_{ii_1}, a_{i_1 i_2}, \ldots, a_{i_q j}$. This is shown most easily as follows. Assume that the grid points are numbered left to right, bottom to top and consider the ith and jth grid points. Then there is a path from the ith point to the jth point consisting of links between grid points as shown in Figure 7.3.2 Each of these links between grid points corresponds to a nonzero element of the matrix A, and it follows that A is irreducible.

Clearly A is diagonally dominant and strict diagonal dominance holds for at least the first and last rows. Therefore, A is irreducibly diagonally dominant and, by Theorem **6.2.6**, is nonsingular. Moreover, since the off-diagonal elements of A are nonpositive while the diagonal elements are

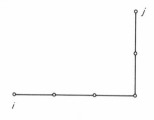

Figure 7.3.2

positive, Theorem 6.2.17 shows that A is an M-matrix. Finally, A is symmetric and, hence, by 6.2.8, positive definite. We summarize these results as follows. The proof of the convergence statements follows exactly as for the matrix of (3).

7.3.1 The matrix A given by (8) and (9) is an irreducibly diagonally dominant M-matrix and is also symmetric positive definite. The Jacobi and Gauss–Seidel iterates converge to the unique solution of the system (10) and, also, the SOR iterates converge for any $\omega \in (0, 2)$.

We note without proof that it can also be shown that the matrix A of (8) is consistently ordered and 2-cyclic, and then Theorem 7.2.9 can be invoked to show that an optimum ω exists for the SOR iteration.

EXERCISES

E7.3.1   Show that the matrix A of (3) is 2-cyclic and consistently ordered.

E7.3.2   Use E7.2.7 to show that, under the condition (4), the eigenvalues of the Jacobi iteration matrix for the matrix A of (3) are real.

E7.3.3   Write out the vector **b** of (10) explicitly.

READING

The SOR theory of Section 7.2 is due to D. M. Young in the early 1950s (see Young [1971]), but the presentation here follows that of Varga [1962]. See also Forsythe and Wasow [1960], Householder [1964], Isaacson and Keller [1966], and Ortega and Rheinboldt [1970].

CHAPTER 8

SYSTEMS OF NONLINEAR EQUATIONS

8.1 LOCAL CONVERGENCE AND RATE OF CONVERGENCE

We consider now the problem of solving the nonlinear system of equations

$$f_i(x_1, \ldots, x_n) = 0, \qquad i = 1, \ldots, n \tag{1}$$

which we usually write in the vector form

$$F\mathbf{x} = \mathbf{0} \tag{2}$$

where $F: R^n \to R^n$ is the mapping† whose components are the f_i; that is, $F\mathbf{x} = (f_1(\mathbf{x}), \ldots, f_n(\mathbf{x}))^{\mathrm{T}}$.

One of the basic iteration procedures for approximating a solution of (2) is **Newton's method**:

$$\mathbf{x}^{k+1} = \mathbf{x}^k - F'(\mathbf{x}^k)^{-1} F\mathbf{x}^k, \, k = 0, 1, \ldots, \tag{3}$$

where $F'(\mathbf{x})$ denotes the Jacobian matrix

$$F'(\mathbf{x}) = \begin{bmatrix} \partial_1 f_1(\mathbf{x}) & \cdots & \partial_n f_1(\mathbf{x}) \\ \vdots & & \vdots \\ \partial_1 f_n(\mathbf{x}) & \cdots & \partial_n f_n(\mathbf{x}) \end{bmatrix}, \tag{4}$$

and where we have used $\partial_i f_j(\mathbf{x})$ to denote the partial derivative of f_j with respect to the ith variable and evaluated at $\mathbf{x}$. In practice, one does not,

---

† For simplicity we will assume that F is defined on all of R^n. This is no essential loss of generality since if F is defined only on a subset D, it can always be extended, in an arbitrary way, to the whole space.

of course, invert $F'(\mathbf{x}^k)$ to carry out (3) but, rather, solves the linear system

$$F'(\mathbf{x}^k)\mathbf{y} = -F\mathbf{x}^k,$$

and adds the "correction" $\mathbf{y}$ to $\mathbf{x}^k$.

In the analysis of Newton's method, it will be necessary to assume that the Jacobian matrix (4) is at least continuous at the solution $\mathbf{x}^*$; that is, $\|F'(\mathbf{x}^* + \mathbf{h}) - F'(\mathbf{x}^*)\| \to 0$ as $\mathbf{h} \to \mathbf{0}$. It is easy to see (E8.1.1) that this will be the case, in any norm, if and only if the partial derivatives $\partial_i f_j$ are all continuous at $\mathbf{x}^*$. On occasion, it will also be useful to assume, instead of continuity of F', that the derivative satisfies the following property.

8.1.1 Definition The mapping $F: R^n \to R^n$ is **(totally** or **Frechet) differentiable** at $\mathbf{x}$ if the Jacobian matix (4) exists at $\mathbf{x}$ and

$$\lim_{\mathbf{h} \to \mathbf{0}} \|F(\mathbf{x} + \mathbf{h}) - F\mathbf{x} - F'(\mathbf{x})\mathbf{h}\| / \|\mathbf{h}\| = 0. \tag{5}$$

Note that if $n = 1$, 8.1.1 reduces to the usual definition of differentiability. Note also that if F is differentiable at $\mathbf{x}$, then F is continuous at $\mathbf{x}$; this follows from the inequality

$$\|F(\mathbf{x} + \mathbf{h}) - F\mathbf{x}\| \le \|F(\mathbf{x} + \mathbf{h}) - F\mathbf{x} - F'(\mathbf{x})\mathbf{h}\| + \|F'(\mathbf{x})\mathbf{h}\|.$$

Finally, we note that it is possible to show† that if the Jacobian matrix is continuous at $\mathbf{x}$, then F is differentiable at $\mathbf{x}$.

One of the basic tools of nonlinear analysis is the mean value theorem. If F is a differentiable function from R^1 to R^1, this states that

$$Fx - Fy = F'(z)(x - y)$$

for some point z between x and y. Unfortunately, this result does not extend verbatim to mappings from R^n to R^n (see E8.1.2). However we are able to prove some results which are often just as useful. For the first, we define the integral‡ $\int_a^b G(t)\,dt$ of a mapping $G: [a, b] \subset R^1 \to R^n$ as the

---

† See, for example, Ortega and Rheinboldt [1970, p. 71].

‡ The integral is assumed to be defined in the Riemann sense; in particular, continuity of the integrand on $[a, b]$ is sufficient for its existence.

vector with components $\int_a^b g_i(t)\, dt$, $i = 1, \ldots, n$, where $g_1, \ldots, g_n$ are the components of G. Thus, for example, if $F: R^n \to R^n$ the relation

$$Fy - Fx = \int_0^1 F'(x + t(y - x))(y - x)\, dt \qquad (6)$$

is equivalent to

$$f_i(y) - f_i(x) = \int_0^1 \sum_{j=1}^n \partial_j f_i(x + t(y - x))(y_j - x_j)\, dt \qquad (7)$$

for $i = 1, \ldots, n$.

For $n = 1$, (6) is simply the fundamental theorem of the integral calculus. Hence the next result is a natural extension of that theorem to n dimensions.

8.1.2   Assume that $F: R^n \to R^n$ is continuously differentiable on a convex† set $D \subset R^n$. Then for any $x, y \in D$, (6) holds.

Proof: For fixed $x, y \in D$ define the functions $g_i: [0, 1] \subset R^1 \to R^1$ by

$$g_i(t) = f_i(x + t(y - x)), \qquad t \in [0, 1], \quad i = 1, \ldots, n.$$

By the convexity of D, it follows that g_i is continuously differentiable on $[0, 1]$ and thus the fundamental theorem of the integral calculus implies that

$$g_i(1) - g_i(0) = \int_0^1 g_i'(t)\, dt. \qquad (8)$$

But a simple calculation shows that

$$g_i'(t) = \sum_{j=1}^n \partial_j f_i(x + t(y - x))(y_j - x_j)$$

so that (8) is equivalent to (7). $\$\$\$$

For the next result, we first need a lemma on integration.

8.1.3   Assume that $G: [a, b] \subset R^1 \to R^n$ is continuous. Then

$$\left\| \int_a^b G(t)\, dt \right\| \leq \int_a^b \| G(t) \|\, dt. \qquad (9)$$

† Recall that a set $D \subset R^n$ is **convex** if $tx + (1 - t)y \in D$ whenever $x, y \in D$ and $t \in [0, 1]$.

Proof: Since any norm is a continuous function† on R^n, both integrals of (9) exist and therefore for any $\varepsilon > 0$ there is a partition $a < t_0 < \cdots < t_p < b$ of $[a, b]$ such that

$$\left\| \int_a^b G(t)\, dt - \sum_{i=1}^p G(t_i)(t_i - t_{i-1}) \right\| < \varepsilon$$

and

$$\left| \int_a^b \|G(t)\|\, dt - \sum_{i=1}^p \|G(t_i)\|(t_i - t_{i-1}) \right| < \varepsilon.$$

Hence

$$\left\| \int_a^b G(t)\, dt \right\| \leq \left\| \sum_{i=1}^p G(t_i)(t_i - t_{i-1}) \right\| + \varepsilon$$

$$\leq \sum_{i=1}^p \|G(t_i)\|(t_i - t_{i-1}) + \varepsilon \leq \int_a^b \|G(t)\|\, dt + 2\varepsilon$$

and, since ε was arbitrary, (9) must be valid.   \$\$\$

By means of 8.1.2 and 8.1.3 we can prove the following useful alternative of the mean value theorem.

8.1.4 Assume that $F: R^n \to R^n$ is continuously‡ differentiable on the convex set $D \subset R^n$. Then for any $\mathbf{x}, \mathbf{y} \in D$,

$$\|F\mathbf{x} - F\mathbf{y}\| \leq \sup_{0 \leq t \leq 1} \|F'(\mathbf{x} + t(\mathbf{y} - \mathbf{x}))\|\, \|\mathbf{y} - \mathbf{x}\|. \tag{10}$$

Proof: By 8.1.2 and 8.1.3 we have

$$\|F\mathbf{x} - F\mathbf{y}\| = \left\| \int_0^1 F'(\mathbf{x} + t(\mathbf{y} - \mathbf{x}))(\mathbf{y} - \mathbf{x})\, dt \right\|$$

$$\leq \int_0^1 \|F'(\mathbf{x} + t(\mathbf{y} - \mathbf{x}))\|\, \|\mathbf{y} - \mathbf{x}\|\, dt$$

$$\leq \sup_{0 \leq t \leq 1} \|F'(\mathbf{x} + t(\mathbf{y} - \mathbf{x}))\| \int_0^1 \|\mathbf{y} - \mathbf{x}\|\, dt$$

which is (10).   \$\$\$

---

† See the proof of Theorem 1.2.4.

‡ We note that this result still holds if F is only differentiable at each point of D, but a different proof is required; see, for example, Ortega and Rheinboldt [1970, p. 69].

Finally, we prove one more often-used estimate.

8.1.5   Assume that $F: R^n \to R^n$ is differentiable and satisfies

$$\|F'(\mathbf{u}) - F'(\mathbf{v})\| \le \gamma \|\mathbf{u} - \mathbf{v}\| \tag{11}$$

for all $\mathbf{u}$, $\mathbf{v}$ in some convex set D. Then for any $\mathbf{x}$, $\mathbf{y} \in D$,

$$\|F\mathbf{y} - F\mathbf{x} - F'(\mathbf{x})(\mathbf{y} - \mathbf{x})\| \le \tfrac{1}{2}\gamma \|\mathbf{x} - \mathbf{y}\|^2.$$

Proof: Since $F'(\mathbf{x})(\mathbf{y} - \mathbf{x})$ is constant with respect to the integration, we have from **8.1.2** that

$$F\mathbf{y} - F\mathbf{x} - F'(\mathbf{x})(\mathbf{y} - \mathbf{x}) = \int_0^1 [F'(\mathbf{x} + t(\mathbf{y} - \mathbf{x})) - F'(\mathbf{x})](\mathbf{y} - \mathbf{x})\, dt.$$

Hence, the result follows by taking norms of both sides, using **8.1.3** and (11), and noting that $\int_0^1 t\, dt = \tfrac{1}{2}$.   \$\$\$

In order to begin our analysis of Newton's method, we consider first the general interation

$$\mathbf{x}^{k+1} = G\mathbf{x}^k, \qquad k = 0, 1, \dots \tag{12}$$

where $G: R^n \to R^n$. A solution of the equation $\mathbf{x} = G\mathbf{x}$ is called a **fixed point** of G, and when G arises as an iteration function for the equation $F\mathbf{x} = \mathbf{0}$, then a solution of $F\mathbf{x} = \mathbf{0}$ should always be a fixed point of G. For example, for Newton's method G is given by

$$G\mathbf{x} \equiv \mathbf{x} - F'(\mathbf{x})^{-1}F\mathbf{x}$$

and, assuming that $F'(\mathbf{x})$ is indeed nonsingular, $\mathbf{x}^*$ is a fixed point of G if and only if $F\mathbf{x}^* = \mathbf{0}$.

In contrast to iterative methods for linear equations, it is usually possible to analyze the convergence of (12) only in a neighborhood about a fixed point.

8.1.6   Definition   A fixed point $\mathbf{x}^*$ of $G: R^n \to R^n$ is a **point of attraction** of the iteration (12) (alternatively, one says that the iteration is **locally convergent** at $\mathbf{x}^*$) if there is an open neighborhood S of $\mathbf{x}^*$ such that whenever $\mathbf{x}^0 \in S$, the iterates (12) are well defined and converge to $\mathbf{x}^*$.

We next give the basic local convergence theorem for the iteration (12).

8.1.7 (Ostrowski's Theorem) Assume that $G: R^n \to R^n$ is differentiable at the fixed point $\mathbf{x}^*$ and that $\rho(G'(\mathbf{x}^*)) < 1$. Then $\mathbf{x}^*$ is a point of attraction of the iteration (12).

Proof: Set $\sigma = \rho(G'(\mathbf{x}^*))$ and take $\varepsilon > 0$. Then, by 1.3.6, there is a norm on R^n such that $\|G'(\mathbf{x}^*)\| \le \sigma + \varepsilon$. Moreover, since G is differentiable, (5) ensures that there is a $\delta > 0$ so that if $S \equiv \{\mathbf{x} : \|\mathbf{x} - \mathbf{x}^*\| < \delta\}$, then

$$\|G\mathbf{x} - G\mathbf{x}^* - G'(\mathbf{x}^*)(\mathbf{x} - \mathbf{x}^*)\| \le \varepsilon \|\mathbf{x} - \mathbf{x}^*\|$$

whenever $\mathbf{x} \in S$. Therefore, for any $\mathbf{x} \in S$

$$\|G\mathbf{x} - G\mathbf{x}^*\| \le \|G\mathbf{x} - G\mathbf{x}^* - G'(\mathbf{x}^*)(\mathbf{x} - \mathbf{x}^*)\| + \|G'(\mathbf{x}^*)(\mathbf{x} - \mathbf{x}^*)\|$$

$$\le (\sigma + 2\varepsilon)\|\mathbf{x} - \mathbf{x}^*\|. \qquad (13)$$

Since $\sigma < 1$, we may assume that $\varepsilon > 0$ is chosen so that $\alpha \equiv \sigma + 2\varepsilon < 1$. Hence, if $\mathbf{x}^0 \in S$, (13) shows that

$$\|\mathbf{x}^1 - \mathbf{x}^*\| = \|G\mathbf{x}^0 - G\mathbf{x}^*\| \le \alpha \|\mathbf{x}^0 - \mathbf{x}^*\|.$$

Therefore, $\mathbf{x}^1 \in S$, and it follows by induction that all $\mathbf{x}^k$ are in S and, moreover, that

$$\|\mathbf{x}^k - \mathbf{x}^*\| \le \alpha \|\mathbf{x}^{k-1} - \mathbf{x}^*\| \le \cdots \le \alpha^k \|\mathbf{x}^0 - \mathbf{x}^*\|.$$

Thus, $\mathbf{x}^k \to \mathbf{x}^*$ as $k \to \infty$. $\$\$\$$

The previous result may be considered to be a local version of Theorem 7.1.1 which showed that the linear iterative process

$$\mathbf{x}^{k+1} = G\mathbf{x}^k \equiv H\mathbf{x}^k + \mathbf{d}, \qquad k = 0, 1, \ldots$$

is convergent for all $\mathbf{x}^0$ if and only if $\rho(H) = \rho(G'(\mathbf{x}^*)) < 1$. Note that in the case of 8.1.7, however, $\rho(G'(\mathbf{x}^*)) < 1$ is not a necessary condition for convergence. (See E8.1.3.)

It is also interesting to note that 8.1.7 is essentially Perron's theorem 4.2.7 on the asymptotic stability of solutions of the perturbed linear difference equation

$$\mathbf{y}^{k+1} = B\mathbf{y}^k + f(\mathbf{y}^k), \qquad k = 0, 1, \ldots. \qquad (14)$$

Indeed, (12) is of the form (14) with $\mathbf{y}^k = \mathbf{x}^k - \mathbf{x}^*$, $B = G'(\mathbf{x}^*)$, and

$$f(\mathbf{y}) = G(\mathbf{x}^* + \mathbf{y}) - G\mathbf{x}^* - G'(\mathbf{x}^*)\mathbf{y}$$

and thus 4.2.7 contains Ostrowski's theorem.

We now apply 8.1.7 to the iterative process

$$\mathbf{x}^{k+1} = \mathbf{x}^k - C(\mathbf{x}^k)^{-1}F\mathbf{x}^k, \qquad k = 0, 1, \dots \tag{15}$$

where C is a given matrix valued function. One concrete example of (15) is Newton's method in which $C(\mathbf{x}) = F'(\mathbf{x})$. Another example is when $C(\mathbf{x})$ is the lower triangular part of $F'(\mathbf{x})$. That is, if we decompose

$$F'(\mathbf{x}) = D(\mathbf{x}) - L(\mathbf{x}) - U(\mathbf{x})$$

into its diagonal, strictly lower triangular, and strictly upper triangular parts, as in Chapter 7, then $C(\mathbf{x}) = D(\mathbf{x}) - L(\mathbf{x})$ and the iteration (15) becomes

$$\mathbf{x}^{k+1} = \mathbf{x}^k - [D(\mathbf{x}^k) - L(\mathbf{x}^k)]^{-1}F\mathbf{x}^k, \qquad k = 0, 1, \dots . \tag{16}$$

In this iteration, only a triangular system of linear equations needs to be solved at each stage, rather than the full system of Newton's method. This is sometimes called the **Newton–Gauss–Seidel** method since (16) reduces to the Gauss–Seidel iteration (7.1.8) when F is linear.

In order to apply 8.1.7 to (15) we need to compute $G'(\mathbf{x}^*)$ where

$$G\mathbf{x} \equiv \mathbf{x} - C(\mathbf{x})^{-1}F\mathbf{x}. \tag{17}$$

Proceeding formally we have

$$G'(\mathbf{x}) = I - C(\mathbf{x})^{-1}F'(\mathbf{x}) - [C(\mathbf{x})^{-1}]'F\mathbf{x} \tag{18}$$

so that, since $F\mathbf{x}^* = \mathbf{0}$,

$$G'(\mathbf{x}^*) = I - C(\mathbf{x}^*)^{-1}F'(\mathbf{x}^*). \tag{19}$$

In order to make this computation rigorous, we would have to discuss the differentiation in (18) more thoroughly. Alternatively, we can obtain (19) as in the next lemma.

8.1.8   Assume that $F: R^n \to R^n$ is differentiable at $\mathbf{x}^*$, where $F\mathbf{x}^* = \mathbf{0}$, and that $C: R^n \to L(R^n)$ is a matrix-valued mapping which is continuous at $\mathbf{x}^*$. Assume, moreover, that $C(\mathbf{x}^*)$ is nonsingular. Then the mapping (17) is well defined in a neighborhood of $\mathbf{x}^*$ and is differentiable at $\mathbf{x}^*$, and (19) is valid.

Proof: We first show that $C(\mathbf{x})$ is nonsingular for all $\mathbf{x}$ in a neighborhood of $\mathbf{x}^*$. Set $\beta = \|C(\mathbf{x}^*)^{-1}\|$ and let ε satisfy $0 < \varepsilon < (2\beta)^{-1}$. By the continuity of C at $\mathbf{x}^*$ there is a $\delta > 0$ so that

$$\|C(\mathbf{x}) - C(\mathbf{x}^*)\| \leq \varepsilon$$

whenever $\mathbf{x} \in S \equiv \{\mathbf{x} : \|\mathbf{x} - \mathbf{x}^*\| \leq \delta\}$. Hence, Theorem 2.1.1 ensures that $C(\mathbf{x})$ is nonsingular and that

$$\|C(\mathbf{x})^{-1}\| \leq \frac{\beta}{1 - \beta\varepsilon} < 2\beta, \qquad \mathbf{x} \in S.$$

Therefore, G is well defined for all $\mathbf{x} \in S$.

Next, since F is differentiable at $\mathbf{x}^*$, we may assume that δ is chosen sufficiently small that

$$\|F\mathbf{x} - F\mathbf{x}^* - F'(\mathbf{x}^*)(\mathbf{x} - \mathbf{x}^*)\| \leq \varepsilon\|\mathbf{x} - \mathbf{x}^*\|$$

for all $\mathbf{x} \in S$. Clearly, $\mathbf{x}^* = G\mathbf{x}^*$ and we can make the estimate

$$\begin{aligned}
\|G\mathbf{x} &- G\mathbf{x}^* - [I - C(\mathbf{x}^*)^{-1}F'(\mathbf{x}^*)](\mathbf{x} - \mathbf{x}^*)\| \\
&= \|C(\mathbf{x}^*)^{-1}F'(\mathbf{x}^*)(\mathbf{x} - \mathbf{x}^*) - C(\mathbf{x})^{-1}F\mathbf{x}\| \\
&\leq \|C(\mathbf{x})^{-1}[F\mathbf{x} - F\mathbf{x}^* - F'(\mathbf{x}^*)(\mathbf{x} - \mathbf{x}^*)]\| \\
&\quad + \|C(\mathbf{x})^{-1}[C(\mathbf{x}^*) - C(\mathbf{x})]C(\mathbf{x}^*)^{-1}F'(\mathbf{x}^*)(\mathbf{x} - \mathbf{x}^*)\| \\
&\leq (2\beta\varepsilon + 2\beta^2\varepsilon\|F'(\mathbf{x}^*)\|)\|\mathbf{x} - \mathbf{x}^*\|
\end{aligned} \tag{20}$$

for all $\mathbf{x} \in S$. Since ε is arbitrary and $\|F'(\mathbf{x}^*)\|$ and β are constant, (20) shows that G is differentiable at $\mathbf{x}^*$ and that (19) holds. $\$\$\$$

We now apply 8.1.8 to the iteration (16).

8.1.9   Assume that $F: R^n \to R^n$ is differentiable in a neighborhood of $\mathbf{x}^*$ where $F\mathbf{x}^* = \mathbf{0}$ and that F' is continuous at $\mathbf{x}^*$. Assume, moreover, that $D(\mathbf{x}^*)$ is nonsingular and that

$$\rho\{[D(\mathbf{x}^*) - L(\mathbf{x}^*)]^{-1}U(\mathbf{x}^*)\} < 1$$

where $U(\mathbf{x})$ is the strictly upper triangular part of $F'(\mathbf{x})$. Then $\mathbf{x}^*$ is a point of attraction of the iteration (16).

Proof: The assumptions imply that $C(\mathbf{x}) \equiv D(\mathbf{x}) - L(\mathbf{x})$ is continuous at $\mathbf{x}^*$ and that $C(\mathbf{x}^*)$ is nonsingular. Hence 8.1.8 ensures that G is differentiable at $\mathbf{x}^*$ and that

$$
\begin{aligned}
G'(\mathbf{x}^*) &= I - [D(\mathbf{x}^*) - L(\mathbf{x}^*)]^{-1} F'(\mathbf{x}^*) \\
&= I - [D(\mathbf{x}^*) - L(\mathbf{x}^*)]^{-1} [D(\mathbf{x}^*) - L(\mathbf{x}^*) - U(\mathbf{x}^*)] \\
&= [D(\mathbf{x}^*) - L(\mathbf{x}^*)]^{-1} U(\mathbf{x}^*).
\end{aligned}
$$

The result then follows from 8.1.7. $\$\$\$$

If $F\mathbf{x} \equiv A\mathbf{x} - \mathbf{b}$, for some $A \in L(R^n)$ with $A = D - L - U$ its decomposition into diagonal, strictly lower, and strictly upper triangular parts, then $F'(\mathbf{x}) = A$ for all $\mathbf{x}$ and the iteration (16) becomes

$$
\mathbf{x}^{k+1} = \mathbf{x}^k - (D - L)^{-1}(A\mathbf{x}^k - \mathbf{b})
$$

which is the Gauss–Seidel iteration discussed in 7.1. The content of 8.1.9 is that the iteration (16) is locally convergent provided that the Gauss–Seidel iteration applied to a linear system with coefficient matrix $F'(\mathbf{x}^*)$ is convergent.

We return now to the Newton iteration (3).

8.1.10   (Local Convergence of Newton's Method)   Assume that $F: R^n \to R^n$ is differentiable at each point of an open neighborhood of a solution $\mathbf{x}^*$ of $F\mathbf{x} = \mathbf{0}$, that F' is continuous at $\mathbf{x}^*$, and that $F'(\mathbf{x}^*)$ is nonsingular. Then $\mathbf{x}^*$ is a point of attraction of the iteration (3) and

$$
\lim_{k \to \infty} \frac{\|\mathbf{x}^{k+1} - \mathbf{x}^*\|}{\|\mathbf{x}^k - \mathbf{x}^*\|} = 0. \tag{21}
$$

Moreover, if

$$
\|F'(\mathbf{x}) - F'(\mathbf{x}^*)\| \le \alpha \|\mathbf{x} - \mathbf{x}^*\| \tag{22}
$$

for all $\mathbf{x}$ in some open neighborhood of $\mathbf{x}^*$, then there is a constant $c < +\infty$ such that

$$
\|\mathbf{x}^{k+1} - \mathbf{x}^*\| \le c \|\mathbf{x}^k - \mathbf{x}^*\|^2 \tag{23}
$$

for all $k \ge k_0$ where k_0 depends on $\mathbf{x}^0$.

Proof: Lemma 8.1.8 ensures that the Newton function

$$
G\mathbf{x} \equiv \mathbf{x} - F'(\mathbf{x})^{-1} F\mathbf{x}
$$

is well defined in a neighborhood of $\mathbf{x}^*$, that G is differentiable at $\mathbf{x}^*$, and that

$$G'(\mathbf{x}^*) = I - F'(\mathbf{x}^*)^{-1}F'(\mathbf{x}^*) = 0. \tag{24}$$

Therefore, it follows from 8.1.7 that $\mathbf{x}^*$ is a point of attraction. Moreover, since G is differentiable at $\mathbf{x}^*$, we must have

$$\lim_{k \to \infty} \frac{\| G\mathbf{x}^k - G\mathbf{x}^* - G'(\mathbf{x}^*)(\mathbf{x}^k - \mathbf{x}^*) \|}{\| \mathbf{x}^k - \mathbf{x}^* \|} = 0$$

whenever $\lim_{k \to \infty} \mathbf{x}^k = \mathbf{x}^*$; by (24), this is equivalent to (21). Finally if (22) holds, then 8.1.5 (with the convex set a line segment emanating from $\mathbf{x}^*$) shows that

$$\| F\mathbf{x} - F\mathbf{x}^* - F'(\mathbf{x}^*)(\mathbf{x} - \mathbf{x}^*) \| \le \tfrac{1}{2}\alpha \| \mathbf{x} - \mathbf{x}^* \|^2.$$

Consequently, the first inequality of the estimate (20) yields

$$\begin{aligned}\| G\mathbf{x} - G\mathbf{x}^* \| &\le \| F'(\mathbf{x})^{-1}[F\mathbf{x} - F\mathbf{x}^* - F'(\mathbf{x}^*)(\mathbf{x} - \mathbf{x}^*)] \| \\ &\quad + \| F'(\mathbf{x})^{-1}[F'(\mathbf{x}) - F'(\mathbf{x}^*)](\mathbf{x} - \mathbf{x}^*) \| \\ &\le \beta\alpha \| \mathbf{x} - \mathbf{x}^* \|^2 + 2\beta\alpha \| \mathbf{x} - \mathbf{x}^* \|^2 \end{aligned}$$

in a suitable neighborhood of $\mathbf{x}^*$. Therefore, (23) holds with $c = 3\beta\alpha$ provided that, for a given $\mathbf{x}^0$, k_0 is chosen so that $\mathbf{x}^k$ lies in this neighborhood for all $k \ge k_0$. $\$\$\$$

The property (21) is known as **superlinear convergence** while (23) is called **quadratic convergence**. We note that (22) is ensured if the component functions f_i of F are all twice continuously differentiable in a neighborhood of $\mathbf{x}^*$. Hence, under these mild differentiability assumptions together with the nonsingularity of $F'(\mathbf{x}^*)$, the content of 8.1.10 is that Newton's method is always locally and quadratically convergent; that is, the Newton iterates *must* converge to $\mathbf{x}^*$, and (23) must hold, as soon as some $\mathbf{x}^k$ is sufficiently close to $\mathbf{x}^*$. The theorem, as is typical of local convergence theorems, thus exhibits an intrinsic property of Newton's method rather than yielding checkable conditions that the iterates will converge starting from a given $\mathbf{x}^0$. This latter type of result will be discussed in the next section.

We end this section by applying 8.1.9 and 8.1.10 to the system of equations

$$f_i(\mathbf{x}) \equiv -x_{i-1} + 2x_i - x_{i+1} + h^2 g(x_i) = 0,$$

$$i = 1, \ldots, n, \quad x_0 = \alpha, \quad x_n = \beta. \tag{25}$$

As discussed previously, this system is a discrete analogue of the two-point boundary value problem

$$y''(t) = g(y(t)), \qquad a \le t \le b, \quad y(a) = \alpha, \quad y(b) = \beta,$$

where α and β are given constants and g is a given function which we shall assume to be twice continuously differentiable and satisfy

$$g'(s) \ge 0, \qquad s \in (-\infty, \infty). \tag{26}$$

Under these assumptions, it is known† that the system (25) has a unique solution $\mathbf{x}^*$.

It is clear that

$$\partial_j f_i(\mathbf{x}) = \begin{cases} -1, & j = i - 1 \\ 2 + h^2 g'(x_i), & j = i \\ -1, & j = i + 1 \end{cases}$$

and 0 otherwise, so that

$$F'(x) = A + \Phi'(\mathbf{x}) \tag{27}$$

where

$$A = \begin{bmatrix} 2 & -1 & \cdot & & \\ -1 & \cdot & \cdot & & \cdot \\ & \cdot & \cdot & \cdot & \\ & & \cdot & \cdot & -1 \\ & & & -1 & 2 \end{bmatrix}$$

$$\Phi'(\mathbf{x}) = h^2 \, \mathrm{diag}(g'(x_1), \dots, g'(x_n)). \tag{28}$$

We have seen previously (6.2.18) that A is an M-matrix, and since, by (26), $\Phi'(\mathbf{x})$ is nonnegative, Theorem 6.2.14 shows that $F'(\mathbf{x})$ is also an M-matrix and hence nonsingular. Since this is true of all $\mathbf{x}$, it is true, in particular, for $\mathbf{x}^*$ and therefore 8.1.10 applies to show that Newton's method is locally convergent. Moreover, if $A = D - L - U$ is the usual decomposition of A into diagonal, strictly lower triangular, and strictly upper triangular parts, then it is a consequence of Theorem 7.1.4 that

$$\rho\{[D + \Phi'(\mathbf{x}^*) - L]^{-1}U\} < 1$$

and therefore, by 8.1.9, the iteration (16) is also locally convergent.

† See, for example, Ortega and Rheinboldt [1970, p. 111].

EXERCISES

E8.1.1   Let $F'(\mathbf{x})$ denote the Jacobian matrix of $F: R^n \to R^n$ at $\mathbf{x}$. Show that $\|F'(\mathbf{y}) - F'(\mathbf{x})\| \to 0$ as $\mathbf{y} \to \mathbf{x}$ if and only if the partial derivatives $\partial_i f_j$ are all continuous at $\mathbf{x}$.

E8.1.2   Let $F: R^2 \to R^2$ be defined by $f_1(\mathbf{x}) = x_1^3$ and $f_2(\mathbf{x}) = x_1^2$. Let $\mathbf{x} = \mathbf{0}$ and $\mathbf{y} = (1, 1)^T$. Show that there is no $\mathbf{z} \in R^2$ such that

$$F\mathbf{x} - F\mathbf{y} = F'(\mathbf{z})(\mathbf{x} - \mathbf{y}).$$

E8.1.3   Consider the iteration $x^{k+1} = Gx^k \equiv x^k - (x^k)^3$, $k = 0, 1, \ldots$, in R^1. Show that 0 is a point of attraction although $G'(0) = 1$. On the other hand, show that 0 is not a point of attraction of $x^{k+1} = Gx^k \equiv x^k + (x^k)^3$, $k = 0, 1$, even though, again, $G'(0) = 1$.

E8.1.4   Consider the iteration

$$\mathbf{x}^{k+1} = \mathbf{x}^k - \omega[D(\mathbf{x}^k) - \omega L(\mathbf{x}^k)]^{-1} F\mathbf{x}^k, \qquad k = 0, 1, \ldots$$

where ω is a fixed parameter. Assume that the conditions of 8.1.9 hold and that

$$\rho\{[D(\mathbf{x}^*) - \omega L(\mathbf{x}^*)]^{-1}[(1 - \omega)D(\mathbf{x}^*) + \omega U(\mathbf{x}^*)]\} < 1.$$

Show that $\mathbf{x}^*$ is a point of attraction.

E8.1.5   Prove 8.1.10 directly by using only the relevant parts of the proof of 8.1.8.

E8.1.6   Consider the Newton iteration in one dimension for the functions $Fx = x^2$ and $Fx = x + x^{1+\alpha}$ where $0 < \alpha < 1$. Show that, in both cases, $x^* = 0$ is a point of attraction of the iteration but that the rate of convergence is not quadratic.

E8.1.7   The iteration

$$\mathbf{x}^{k+1} = \mathbf{x}^k - F'(\mathbf{x}^0)^{-1} F\mathbf{x}^k, \qquad k = 0, 1, \ldots$$

is known as the **simplified Newton iteration**. Use 8.1.7 to state a local convergence theorem for this iteration.

E8.1.8   Show that the unique solution $\mathbf{x}^*$ of the system

$$f_i(\mathbf{x}) \equiv -x_{i-1} + 2x_i - x_{i+1} + x_i^3, \qquad i = 1, \ldots, n; \quad x_0 = \alpha, \quad x_{n+1} = \beta$$

is a point of attraction for Newton's method.

E8.1.9   The method

$$\mathbf{x}^{k+1} = \mathbf{x}^k - D(\mathbf{x}^k)^{-1}F\mathbf{x}^k, \qquad k = 0, 1, \ldots,$$

where $D(\mathbf{x})$ is the diagonal matrix consisting of the diagonal elements of $F'(\mathbf{x})$, is sometimes called the **Newton–Jacobi method**. Show that this method reduces to that of equations (3) and (4) of the Introduction. State and prove a result analogous to **8.1.9**.

8.2   ERROR ESTIMATES

The local convergence results of the previous section serve to establish certain intrinsic properties of the iterative methods considered. However, they are generally useless when one tries to ascertain whether an iterative process will converge starting from a given $\mathbf{x}^0$, and, also, what is the error if the process is stopped with the kth iterate. In this section we will give several results which are at least potentially useful in this regard.

A special case of the basic convergence theorem **7.1.1** for the linear iterative process

$$\mathbf{x}^{k+1} = H\mathbf{x}^k + \mathbf{d}, \qquad k = 0, 1, \ldots \tag{1}$$

is that $\|H\| < 1$ is a sufficient condition in order that these iterates converge to the unique solution of $\mathbf{x} = H\mathbf{x} + \mathbf{d}$ for any $\mathbf{x}^0$. We shall extend this result to the nonlinear iteration

$$\mathbf{x}^{k+1} = G\mathbf{x}^k, \qquad k = 0, 1, \ldots \tag{2}$$

in the following way.

8.2.1   Definition   The mapping $G: R^n \to R^n$ is a **contraction** on the set $D \subset R^n$ if there is a constant $\alpha < 1$ so that

$$\|G\mathbf{x} - G\mathbf{y}\| \le \alpha \|\mathbf{x} - \mathbf{y}\| \tag{3}$$

for all $\mathbf{x}, \mathbf{y}$ in D.

Clearly, the mapping $G\mathbf{x} \equiv H\mathbf{x} + \mathbf{d}$ is a contraction on all of R^n if $\|H\| < 1$; in this case, $\alpha = \|H\|$. (See also **E8.2.1**.)

We now prove one of the most famous theorems of analysis. In the sequel, the notation $GD \subset D$ will mean that $G\mathbf{x} \in D$ whenever $\mathbf{x} \in D$.

8.2.2 (Contraction Mapping Theorem)   Assume that $G: R^n \to R^n$ is a contraction on a closed set D and that $GD \subset D$. Then G has a unique fixed point $\mathbf{x}^* \in D$ and for any $\mathbf{x}^0 \in D$, the iterates (2) converge to $\mathbf{x}^*$. Moreover,

$$\|\mathbf{x}^k - \mathbf{x}^*\| \leq \alpha(1 - \alpha)^{-1}\|\mathbf{x}^k - \mathbf{x}^{k-1}\|, \qquad k = 1, 2, \ldots \tag{4}$$

where α is the contraction constant of (3).

Proof: By the assumption that $GD \subset D$, all of the iterates $\mathbf{x}^k$ lie in D and thus by (3)

$$\|\mathbf{x}^{k+1} - \mathbf{x}^k\| = \|G\mathbf{x}^k - G\mathbf{x}^{k-1}\| \leq \alpha\|\mathbf{x}^k - \mathbf{x}^{k-1}\|, \qquad k = 1, 2, \ldots.$$

Therefore, for any $k \geq 0$ and $p \geq 1$

$$\|\mathbf{x}^{k+p} - \mathbf{x}^k\| \leq \sum_{i=1}^{p} \|\mathbf{x}^{k+i} - \mathbf{x}^{k+i-1}\| \leq (\alpha^p + \cdots + \alpha)\|\mathbf{x}^k - \mathbf{x}^{k-1}\|$$

$$\leq \frac{\alpha^k}{1 - \alpha}\|\mathbf{x}^1 - \mathbf{x}^0\|. \tag{5}$$

This shows that the sequence $\{\mathbf{x}^k\}$ is a Cauchy sequence and hence, because D is closed, has a limit $\mathbf{x}^* \in D$. Since G is a contraction, it is continuous and therefore $\lim_{k \to \infty} G\mathbf{x}^k = G\mathbf{x}^*$, which, by (2), also shows that $\mathbf{x}^*$ is a fixed point of G. The uniqueness of $\mathbf{x}^*$ is apparent from the inequality

$$\|\mathbf{x}^* - \mathbf{y}^*\| = \|G\mathbf{x}^* - G\mathbf{y}^*\| \leq \alpha\|\mathbf{x}^* - \mathbf{y}^*\|$$

which is absurd if the fixed point $\mathbf{y}^* \in D$ is distinct from $\mathbf{x}^*$. Finally, the estimate (4) follows from the first two inequalities of (5) by letting p tend to $+\infty$. $\$\$\$$

The error estimate (4) is extremely useful when it can be applied. Suppose that we know that $\alpha = 0.9$ and that we are willing to tolerate an error of 10^{-4} in the solution; that is, we wish to guarantee that the approximate solution $\mathbf{x}^k$ satisfies $\|\mathbf{x}^* - \mathbf{x}^k\| < 10^{-4}$. We can ensure this provided that we iterate until $\|\mathbf{x}^k - \mathbf{x}^{k-1}\| \leq 10^{-5}$ since (4) then shows that

$$\|\mathbf{x}^k - \mathbf{x}^*\| \leq 9\|\mathbf{x}^k - \mathbf{x}^{k-1}\| < 10^{-4}.$$

The main difficulty in applying 8.2.2 is to obtain an estimate for the contraction constant. Sometimes this may be done by estimating the derivative as shown in the next result, which is an immediate consequence of the mean value theorem 8.1.4.

8.2.3 Suppose that $G: R^n \to R^n$ is continuously differentiable on the convex set D and that

$$\|G'(\mathbf{x})\| \le \alpha \tag{6}$$

for all $\mathbf{x} \in D$. Then (3) holds for all $\mathbf{x}, \mathbf{y}$ in D.

As an example of the use of these results, we consider the system (8.1.25) written in the form

$$F\mathbf{x} \equiv A\mathbf{x} + \Phi\mathbf{x} \tag{7}$$

where

$$A = \begin{bmatrix} 2 & -1 & & & \mathbf{O} \\ -1 & 2 & \ddots & & \\ & \ddots & \ddots & \ddots & \\ & & \ddots & \ddots & -1 \\ \mathbf{O} & & & -1 & 2 \end{bmatrix}, \qquad \Phi\mathbf{x} = \begin{bmatrix} h^2 g(x_1) + \alpha \\ h^2 g(x_2) \\ \vdots \\ h^2 g(x_{n-1}) \\ h^2 g(x_n) + \beta \end{bmatrix}. \tag{8}$$

We will assume that

$$0 \le g'(s) \le \beta, \qquad -\infty < s < +\infty \tag{9}$$

and consider the "Picard iteration"

$$\mathbf{x}^{k+1} = (A + \gamma I)^{-1}(\gamma\mathbf{x}^k - \Phi\mathbf{x}^k), \qquad k = 0, 1, \ldots \tag{10}$$

for some suitable constant γ.

8.2.4 Assume that $A \in L(R^n)$ and $\Phi: R^n \to R^n$ are defined by (8) where $g: R^1 \to R^1$ is continuously differentiable and satisfies (9). Then the system (7) has a unique solution $\mathbf{x}^*$ and the iterates $\mathbf{x}^k$ of (10) with $\gamma = h^2\beta/2$ are well defined and converge to $\mathbf{x}^*$ for any $\mathbf{x}^0$.

Proof: Define $G: R^n \to R^n$ by

$$G\mathbf{x} \equiv (A + \gamma I)^{-1}(\gamma\mathbf{x} - \Phi\mathbf{x}). \tag{11}$$

By 6.2.15 and 6.2.18 A is positive definite and, since $\gamma > 0$, $A + \gamma I$ is also positive definite. Hence, G is well defined and has a fixed point $\mathbf{x}^*$ if and only if $\mathbf{x}^*$ is a solution of (7). Moreover, G is continuously differentiable on R^n and

$$\|G'(\mathbf{x})\|_2 = \|(A + \gamma I)^{-1}[\gamma I - \Phi'(\mathbf{x})]\|_2 \leq \frac{\gamma}{\lambda + \gamma} < 1$$

where $\lambda > 0$ is the minimum eigenvalue of A. Therefore, **8.2.3** ensures that G is a contraction on all of R^n and the result then follows from **8.2.2**.   \$\$\$

If G is a contraction on only part of R^n, our results so far do not answer the question of whether for a given $\mathbf{x}^0$ the iterates will converge. We discuss next a potentially useful result in this regard.

8.2.5   Let $G: R^n \to R^n$ be a contraction on D with constant α and assume that $\mathbf{x}^0 \in D$ is such that

$$S \equiv \left\{ \mathbf{x} : \|\mathbf{x} - G\mathbf{x}^0\| \leq \frac{\alpha}{1 - \alpha} \|G\mathbf{x}^0 - \mathbf{x}^0\| \right\} \subset D. \tag{12}$$

Then the iterates (2) converge to the unique fixed point $\mathbf{x}^*$ of G in S.

Proof: If $\mathbf{x} \in S$, then

$$\|G\mathbf{x} - G\mathbf{x}^0\| \leq \alpha \|\mathbf{x} - \mathbf{x}^0\| \leq \alpha[\|\mathbf{x} - G\mathbf{x}^0\| + \|G\mathbf{x}^0 - \mathbf{x}^0\|]$$

$$\leq \left(\frac{\alpha^2}{1 - \alpha} + \alpha \right) \|G\mathbf{x}^0 - \mathbf{x}^0\| = \frac{\alpha}{1 - \alpha} \|G\mathbf{x}^0 - \mathbf{x}^0\|.$$

Hence $G\mathbf{x} \in S$ so that $GS \subset S$. Clearly, S is closed so that **8.2.2** applies.   \$\$\$

It is possible to apply the previous result to Newton's method, but the resulting theorem is not the best possible. We will prove the following sharper result.

8.2.6   (Newton–Kantorovich Theorem)   Assume that $F: R^n \to R^n$ is differentiable on a convex set D and that

$$\|F'(\mathbf{x}) - F'(\mathbf{y})\| \leq \gamma \|\mathbf{x} - \mathbf{y}\| \tag{13}$$

for all $\mathbf{x}, \mathbf{y} \in D$. Suppose that there is an $\mathbf{x}^0 \in D$ such that

$$\|F'(\mathbf{x}^0)^{-1}\| \leq \beta, \qquad \|F'(\mathbf{x}^0)^{-1}F\mathbf{x}^0\| = \eta, \tag{14}$$

and

$$\alpha \equiv \beta\gamma\eta < \tfrac{1}{2}. \tag{15}$$

Assume that

$$S \equiv \{x : \| x - x^0 \| \le t_* \} \subset D, \qquad t_* = (1/\beta\gamma)[1 - (1 - 2\alpha)^{1/2}]. \tag{16}$$

Then the Newton iterates

$$x^{k+1} = x^k - F'(x^k)^{-1}Fx^k, \qquad k = 0, 1, \dots \tag{17}$$

are well defined and converge to a solution† x^* of $Fx = 0$ in S. Moreover, we have the error estimate

$$\| x^* - x^1 \| \le 2\beta\gamma \| x^1 - x^0 \|^2. \tag{18}$$

Proof: Since, by (13),

$$\| F'(x) - F'(x^0) \| \le \gamma \| x - x^0 \| \le \gamma t_* < 1/\beta$$

for any $x \in S$, the perturbation lemma 2.1.1 shows that $F'(x)$ is nonsingular and that

$$\| F'(x)^{-1} \| \le \frac{\beta}{1 - \beta\gamma \| x - x^0 \|}, \qquad x \in S. \tag{19}$$

Consequently, the Newton function

$$Gx = x - F'(x)^{-1}Fx$$

is well defined on S and if x and Gx are in S then

$$\| G(Gx) - Gx \| = \| F'(Gx)^{-1}F(Gx) \| \le \frac{\beta \| F(Gx) \|}{1 - \beta\gamma \| x^0 - Gx \|}, \qquad x, Gx \in S. \tag{20}$$

But, by 8.1.5 and the definition of G, we have

$$\| F(Gx) \| = \| F(Gx) - Fx - F'(x)(Gx - x) \| \le \tfrac{1}{2}\gamma \| Gx - x \|^2$$

so that (20) becomes

$$\| G(Gx) - Gx \| \le \frac{\beta\gamma \| Gx - x \|^2}{2(1 - \beta\gamma \| x^0 - Gx \|)}, \qquad x, Gx \in S. \tag{21}$$

We shall return to this basic estimate directly, but first we define the scalar sequence

$$t_{k+1} = t_k - \frac{\tfrac{1}{2}\beta\gamma t_k^2 - t_k + \eta}{\beta\gamma t_k - 1}, \qquad k = 0, 1, \dots, \quad t_0 = 0. \tag{22}$$

---

† It may be shown that x^* is not only unique in S but in the larger ball centered at x^0 with radius $(1/\beta\gamma)[1 + (1 - 2\alpha)^{1/2}]$. See Ortega and Rheinboldt [1970].

It is easy to show that the sequence $\{t_k\}$ is well defined, monotonically increasing, and converges to t_*. In fact, these properties are geometrically obvious (see Figure 8.2.1) by noting that (22) is just Newton's method for the quadratic $p(t) \equiv \frac{1}{2}\beta\gamma t^2 - t + \eta$, whose smaller root is t_*.

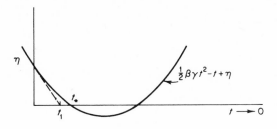

Figure 8.2.1

Now assume that $\mathbf{x}^0, \ldots, \mathbf{x}^k$ are in S and that

$$\|\mathbf{x}^i - \mathbf{x}^{i-1}\| \le t_i - t_{i-1} \tag{23}$$

for $i = 1, \ldots, k$; for $k = 1$, this is trivial since $\|\mathbf{x}^1 - \mathbf{x}^0\| = \|F'(\mathbf{x}^0)^{-1}F\mathbf{x}^0\| = \eta = t_1$. By (23),

$$\|\mathbf{x}^k - \mathbf{x}^0\| \le \sum_{i=1}^{k} \|\mathbf{x}^i - \mathbf{x}^{i-1}\| \le \sum_{i=1}^{k} (t_i - t_{i-1}) = t_k$$

and then (21) shows that

$$\|\mathbf{x}^{k+1} - \mathbf{x}^k\| = \|G(G\mathbf{x}^{k-1}) - G\mathbf{x}^{k-1}\|$$

$$\le \frac{\beta\gamma \|\mathbf{x}^k - \mathbf{x}^{k-1}\|^2}{2(1 - \beta\gamma\|\mathbf{x}^0 - \mathbf{x}^k\|)}$$

$$\le \frac{\beta\gamma(t_k - t_{k-1})^2}{2(1 - \beta\gamma t_k)} = t_{k+1} - t_k \tag{24}$$

where the last equality follows from the fact that

$$t_{k+1} - t_k = \frac{-p(t_k)}{p'(t_k)}$$

$$= \frac{-1}{\beta\gamma t_k - 1}\left[p(t_{k-1}) + p'(t_{k-1})(t_k - t_{k-1}) + \frac{1}{2}p''(t_{k-1})(t_k - t_{k-1})^2\right]$$

$$= \frac{\beta\gamma(t_k - t_{k-1})^2}{2(1 - \beta\gamma t_k)}.$$

It now follows from (24) that $\|\mathbf{x}^{k+1} - \mathbf{x}^0\| \le t_{k+1} \le t_*$ so that $\mathbf{x}^{k+1} \in S$.

Hence the inductive step is complete, and we have shown that $\mathbf{x}^i \in S$ and (23) holds for all i. Therefore, for any $k \geq 0$ and $p \geq 1$

$$\|\mathbf{x}^{k+p} - \mathbf{x}^k\| \leq \sum_{i=1}^{p} (t_{k+i} - t_{k+i-1}) = t_{k+p} - t_k, \qquad (25)$$

and, since $t_k \to t_*$ as $k \to \infty$, $\{t_k\}$ is a Cauchy sequence. Therefore (25) shows that $\{\mathbf{x}^k\}$ is also a Cauchy sequence, which, by the closure of S, must have a limit $\mathbf{x}^* \in S$. But

$$\|F\mathbf{x}^k\| = \|F'(\mathbf{x}^k)(\mathbf{x}^{k+1} - \mathbf{x}^k)\| \leq \|F'(\mathbf{x}^k)\| \, \|\mathbf{x}^{k+1} - \mathbf{x}^k\| \qquad (26)$$

and since $\|\mathbf{x}^{k+1} - \mathbf{x}^k\| \to 0$ as $k \to \infty$ and $\|F'(\mathbf{x})\|$ is bounded on S the right-hand side of (26) tends to zero as $k \to \infty$. Hence $F\mathbf{x}^k \to \mathbf{0}$ as $k \to \infty$, and the continuity of F implies that $F\mathbf{x}^* = \mathbf{0}$.

Finally, to obtain the estimate (18), note that it follows from (25) as $p \to \infty$ that

$$\|\mathbf{x}^* - \mathbf{x}^k\| \leq t_* - t_k.$$

In particular, we have

$$\|\mathbf{x}^* - \mathbf{x}^1\| \leq t_* - \eta = \tfrac{1}{2}\beta\gamma t_*^2 \leq 2\beta\gamma\eta^2 = 2\beta\gamma\|\mathbf{x}^1 - \mathbf{x}^0\|^2,$$

where we have used the inequality $1 - \sqrt{1 - \delta} \leq \delta$ to estimate t_*. $\$\$\$$

Note that the previous result shows not only that the Newton iterates starting from a given $\mathbf{x}^0$ will converge but also that a solution exists. In addition, the estimate (18) is another manifestation of the quadratic convergence property of Newton's method. The conditions of the theorem are usually very difficult to verify in practice, however; in particular, there must usually be a delicate balance between the set D being sufficiently large so that (16) will hold and at the same time being sufficiently small so that the quantity γ is as small as possible. Clearly, the ease with which these conditions may be satisfied depends crucially on how close $\mathbf{x}^0$ is to a solution, as reflected by the quantity $\|F'(\mathbf{x}^0)^{-1}F\mathbf{x}^0\|$.

The importance of 8.2.6 lies perhaps more in the direction of obtaining a posteriori error estimates† than in guaranteeing the convergence of the Newton iterates. That is, suppose that the Newton iterates are computed

---

† The error estimate of this theorem is usually stated in the form $\|\mathbf{x}^* - \mathbf{x}^k\| \leq (2\alpha)^{2^k}/(\beta\gamma 2^k)$, $k = 0, 1, \ldots$; for a proof of this inequality, see Ortega and Rheinboldt [1970, p. 423]. We have given an estimate of the form (18) since its proof is immediate and it is just as useful for a posteriori bounds.

until an error criterion of the form $\|\mathbf{x}^j - \mathbf{x}^{j-1}\| \leq \varepsilon$ is satisfied, and then $\mathbf{x}^j$ is taken as the approximate solution. In order to obtain an estimate on the accuracy of $\mathbf{x}^j$ we apply 8.2.6 with $\mathbf{x}^0$ replaced by $\mathbf{x}^{j-1}$. Then $\alpha \leq \beta\gamma\varepsilon$, where $\beta \geq \|F'(\mathbf{x}^{j-1})^{-1}\|$, and provided that the conditions of the theorem can be satisfied, (18) ensures that the estimate $\|\mathbf{x}^* - \mathbf{x}^j\| \leq 2\beta\gamma\varepsilon^2$ holds.

We end this section with a short discussion of a device which is sometimes useful in widening the domain of convergence of a given method, or, alternatively, as a procedure to obtain sufficiently close starting points for the method.

Given the mapping $F: R^n \to R^n$, we define a one-parameter family of equations

$$H(\mathbf{x}, t) = \mathbf{0}, \qquad t \in [0, 1] \tag{27}$$

with the properties that $H(\mathbf{x}, 1) = F\mathbf{x}$ and that the equation $H(\mathbf{x}, 0) = \mathbf{0}$ has a known solution $\mathbf{x}^0$. For example, we may take

$$H(\mathbf{x}, t) \equiv tF\mathbf{x} + (1 - t)F_0\mathbf{x} \tag{28}$$

where, for a given $\mathbf{x}^0$, $F_0\mathbf{x} \equiv F\mathbf{x} - F\mathbf{x}^0$ or $F_0\mathbf{x} \equiv \mathbf{x} - \mathbf{x}^0$.

Now suppose that for each $t \in [0, 1]$ the equation (27) has a solution $\mathbf{x}(t)$ which depends continuously on t; that is, suppose there is a continuous function $\mathbf{x}: [0, 1] \to R^n$ such that

$$H(\mathbf{x}(t), t) = \mathbf{0}, \qquad t \in [0, 1], \quad \mathbf{x}(0) = \mathbf{x}^0. \tag{29}$$

Then $\mathbf{x}$ describes a space curve in R^n with one endpoint at $\mathbf{x}^0$ and the other endpoint at the solution $\mathbf{x}^*$ of $F\mathbf{x} = \mathbf{0}$; this is depicted in Figure 8.2.2.

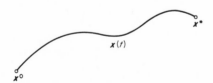

Figure 8.2.2

One way of using (29) to obtain an approximation to $\mathbf{x}^*$ is as follows. Partition the interval $[0, 1]$ by the grid points

$$0 = t_0 < t_1 < \cdots < t_N = 1$$

and solve the equations

$$H(\mathbf{x}, t_i) = \mathbf{0}, \qquad i = 1, \ldots, N$$

by some iterative method, for example, Newton's method, which uses the solution x^{i-1} of the $(i-1)$st problem as a starting approximation for the ith problem. If $t_i - t_{i-1}$ is suitably small, then x^{i-1} will be sufficiently close to x^i so that convergence will occur.

We next consider a somewhat different approach and for simplicity we restrict ourselves to the imbedding

$$Fx(t) - (1 - t)Fx^0 = 0.$$

If we differentiate this equation with respect to t, we obtain

$$F'\big(x(t)\big)x'(t) + Fx^0 = 0,$$

so that, assuming that $F'(x)$ is nonsingular along the solution curve, we have

$$x'(t) = -F'\big(x(t)\big)^{-1}Fx^0. \tag{30}$$

Therefore, the solution curve $x(t)$ is the solution of the differential equation (30) with initial condition $x(0) = x^0$. This is **Davidenko's method of differentiation with respect to a parameter**. In principle, we can approximate a solution of (30) by any of the initial value methods of Chapter 5.

EXERCISES

E8.2.1 Define $G: R^n \to R^n$ by $Gx = Hx + d$ where $H \in L(R^n)$ and $d \in R^n$. Show that there is a norm on R^n in which G is a contraction if and only if $\rho(H) < 1$.

E8.2.2 Suppose that $G: R^n \to R^n$ has the property that for any closed bounded set $C \in R^n$ there is a constant $\alpha_C < 1$ such that

$$\|Gx - Gy\| \le \alpha_C\|x - y\|$$

for all $x, y \in C$, and assume that G has a fixed point x^*. Show that x^* is unique and that the iterates $x^{k+1} = Gx^k$, $k = 0, 1, \ldots$ converge to x^* for any x^0. Give an example in R^1 which shows that the condition on G is not sufficient to guarantee that G has a fixed point.

E8.2.3 Let $F: R^n \to R^n$ be defined by (7) and (8) and assume that, in place of (9), $g'(s) \ge 0$ for all $s \in (-\infty, \infty)$. Show that the function G of (11) is not necessarily a contraction on all of R^n no matter what (fixed) value of γ is chosen.

E8.2.4 Assume that $F: R^n \to R^n$ is continuously differentiable in a neighborhood of a solution $\mathbf{x}^*$ of $F\mathbf{x} = 0$ and that $F'(\mathbf{x}^*)$ is nonsingular. Show that the Newton function $G\mathbf{x} = \mathbf{x} - F'(\mathbf{x})^{-1}F\mathbf{x}$ is a contraction in some neighborhood of $\mathbf{x}^*$.

8.3 GLOBAL CONVERGENCE

We turn now to the problem of global convergence of iterative methods, that is, convergence for any $\mathbf{x}^0$. The results of Chapter 7 for linear equations were all of this type, but we have seen in the previous two sections that for nonlinear equations the situation is quite different and global convergence theorems are the exception rather than the rule.

Perhaps the simplest global convergence theorem is a special case of the contraction mapping theorem **8.2.2**.

8.3.1 Assume that $G: R^n \to R^n$ is a contraction on all of R^n. Then G has a unique fixed point $\mathbf{x}^*$ and the iterates

$$\mathbf{x}^{k+1} = G\mathbf{x}^k, \qquad k = 0, 1, \ldots \tag{1}$$

converge to $\mathbf{x}^*$ for any $\mathbf{x}^0$.

We already gave an example of the use of **8.3.1** in **8.2.4**. However, it is usually difficult, especially without very stringent assumptions, to show that a given iteration function G is a contraction on all of R^n. In particular, the application of **8.3.1** to Newton's method yields rather limited results. Therefore, we will pursue a different course in this section and extend to n dimensions the following well-known result for Newton's method in one dimension.

We assume that $F: R^1 \to R^1$ is convex and monotone increasing. Then for any x^0 to the right of the root x^* the Newton iterates will converge monotonically down to x^*, while if x^0 is to the left of x^* the first Newton iterate will be to the right of x^* so that the succeeding iterates will converge monotonically downward. The situation is depicted in Figure 8.3.1.

In order to extend this geometrically obvious result to n dimensions, we need to extend the concept of a convex function to mappings $F: R^n \to R^n$. In the sequel, we shall use the natural partial ordering for vectors and matrices introduced in **6.2**; that is, $\mathbf{x} \le \mathbf{y}$ if $x_i \le y_i$, $i = 1, \ldots, n$, and $A \le B$ if $a_{ij} \le b_{ij}$, $i, j = 1, \ldots, n$.

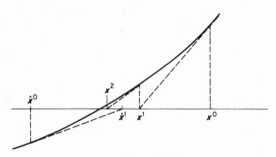

Figure 8.3.1

8.3.2   Definition   The mapping $F: R^n \to R^n$ is **convex** on a convex set $D \subset R^n$ if

$$F(\alpha\mathbf{x} + (1 - \alpha)\mathbf{y}) \le \alpha F\mathbf{x} + (1 - \alpha)F\mathbf{y} \qquad (2)$$

for all $\mathbf{x}, \mathbf{y} \in D$ and $\alpha \in [0, 1]$.

Note that when $n = 1$, 8.3.2 reduces to the usual definition of a convex function.

Recall that a one-dimensional differentiable function is convex if and only if the graph of the function lies above all its tangent lines. This extends to n dimensions in terms of the following basic **differential inequality**.

8.3.3   Assume that $F: R^n \to R^n$ is differentiable on the convex set D. Then F is convex on D if and only if

$$F\mathbf{y} - F\mathbf{x} \ge F'(\mathbf{x})(\mathbf{y} - \mathbf{x}) \qquad (3)$$

for all $\mathbf{x}, \mathbf{y} \in D$.

Proof: Suppose first that (3) holds for all $\mathbf{x}, \mathbf{y} \in D$, and for given $\mathbf{u}, \mathbf{v} \in D$ and $\alpha \in [0, 1]$, set $\mathbf{z} = \alpha\mathbf{u} + (1 - \alpha)\mathbf{v}$. Then $\mathbf{z} \in D$ and hence

$$F\mathbf{u} - F\mathbf{z} \ge F'(\mathbf{z})(\mathbf{u} - \mathbf{z}), \qquad F\mathbf{v} - F\mathbf{z} \ge F'(\mathbf{z})(\mathbf{v} - \mathbf{z}).$$

If we multiply these inequalities by α and $1 - \alpha$, respectively, and then add we obtain

$$\alpha F\mathbf{u} + (1 - \alpha)F\mathbf{v} - F\mathbf{z} \ge F'(\mathbf{z})[\alpha\mathbf{u} + (1 - \alpha)\mathbf{v} - \mathbf{z}] = \mathbf{0}.$$

Hence F is convex on D. Conversely, if F is convex on D, then for any $\mathbf{x}, \mathbf{y} \in D$ and $\alpha \in (0, 1]$ we have, by rearranging (2),

$$F\mathbf{y} - F\mathbf{x} \geq (1/\alpha)[F(\mathbf{x} + \alpha(\mathbf{y} - \mathbf{x})) - F\mathbf{x}].$$

Since F is differentiable at $\mathbf{x}$ the right-hand side of this inequality tends to $F'(\mathbf{x})(\mathbf{y} - \mathbf{x})$ as $\alpha \to 0$, which shows that (3) is valid. $\$\$\$$

We are now able to prove the global convergence theorem previously mentioned. In the following, the notation $A \geq 0$ will mean, as usual, that the matrix $A \in L(R^n)$ has nonnegative elements.

8.3.4   (Newton–Baluev Theorem†)   Assume that $F: R^n \to R^n$ is continuously differentiable and convex on all of R^n, that $F'(\mathbf{x})$ is nonsingular and $F'(\mathbf{x})^{-1} \geq 0$ for all $\mathbf{x} \in R^n$, and that $F\mathbf{x} = \mathbf{0}$ has a solution $\mathbf{x}^*$. Then $\mathbf{x}^*$ is unique and the Newton iterates

$$\mathbf{x}^{k+1} = \mathbf{x}^k - F'(\mathbf{x}^k)^{-1}F\mathbf{x}^k, \qquad k = 0, 1, \ldots \tag{4}$$

converge to $\mathbf{x}^*$ for any $\mathbf{x}^0$. Moreover,

$$\mathbf{x}^* \leq \mathbf{x}^{k+1} \leq \mathbf{x}^k, \qquad k = 1, 2, \ldots. \tag{5}$$

Proof: For arbitrary $\mathbf{x}^0 \in R^n$, **8.3.3** and the definition of $\mathbf{x}^1$ show that

$$F\mathbf{x}^1 - F\mathbf{x}^0 \geq F'(\mathbf{x}^0)(\mathbf{x}^1 - \mathbf{x}^0) = -F\mathbf{x}^0.$$

Hence $F\mathbf{x}^1 \geq 0$. Another application of **8.3.3** yields

$$\mathbf{0} = F\mathbf{x}^* \geq F\mathbf{x}^1 + F'(\mathbf{x}^1)(\mathbf{x}^* - \mathbf{x}^1)$$

and, since $F'(\mathbf{x}^1)^{-1} \geq \mathbf{0}$, we may multiply both sides of this inequality by $F'(\mathbf{x}^1)^{-1}$ in order to obtain

$$\mathbf{x}^* \leq \mathbf{x}^1 - F'(\mathbf{x}^1)^{-1}F\mathbf{x}^1 \leq \mathbf{x}^1.$$

By induction, we may prove in a completely analogous way that

$$F\mathbf{x}^k \geq 0, \qquad \mathbf{x}^k \geq x^*, \qquad k = 1, 2, \ldots. \tag{6}$$

Hence,

$$\mathbf{x}^{k+1} = \mathbf{x}^k - F'(\mathbf{x}^k)^{-1}F\mathbf{x}^k \leq \mathbf{x}^k, \qquad k = 1, 2, \ldots$$

which, together with (6), shows that (5) is valid. Therefore each component

† Apparently first proved by the Russian mathematician A. Baluev, *Dokl. Akad. Nauk SSSR* **83** (1952), 781–784.

sequence $\{x_i^k\}$, $i = 1, \ldots, n$ of $\{\mathbf{x}^k\}$ is monotone decreasing and bounded below by x_i^* and thus has a limit y_i; that is, the vector sequence $\{\mathbf{x}^k\}$ has a limit $\mathbf{y}$. From the continuity of F' it follows that

$$F\mathbf{y} = \lim_{k \to \infty} F\mathbf{x}^k = \lim_{k \to \infty} F'(\mathbf{x}^k)(\mathbf{x}^k - \mathbf{x}^{k+1}) = 0.$$

Now suppose that $\mathbf{x}^*$ and $\mathbf{y}^*$ are any two solutions of $F\mathbf{x} = \mathbf{0}$. Then, using 8.3.3, we have

$$0 = F\mathbf{y}^* - F\mathbf{x}^* \geq F'(\mathbf{x}^*)(\mathbf{y}^* - \mathbf{x}^*)$$

so that if we multiply through by $F'(\mathbf{x}^*)^{-1}$ we obtain $\mathbf{y}^* \leq \mathbf{x}^*$. It follows by reversing the roles of $\mathbf{x}^*$ and $\mathbf{y}^*$ that $\mathbf{x}^* \leq \mathbf{y}^*$ and therefore $\mathbf{x}^* = \mathbf{y}^*$. Hence, $\mathbf{x}^*$ is the unique solution and therefore equal to the limit vector $\mathbf{y}$. $\$\$\$$

The inequality (5) is very useful because if we stop the iteration at any k we know that $\mathbf{x}^k$ is an "upper bound" for the solution $\mathbf{x}^*$. In fact, by means of a second sequence

$$\hat{\mathbf{x}}^{k+1} = \hat{\mathbf{x}}^k - F'(\mathbf{x}^k)^{-1}F\hat{\mathbf{x}}^k, \qquad k = 1, 2, \ldots \tag{7}$$

we may obtain lower bounds, and hence the inequality

$$\hat{\mathbf{x}}^k \leq \mathbf{x}^* \leq \mathbf{x}^k, \qquad k = 1, 2, \ldots, \tag{8}$$

which provides the basis for an excellent stopping criterion for the iteration. The situation is depicted for $n = 1$ in Figure 8.3.2. Note that F' is evaluated in (7) at the Newton iterate $\mathbf{x}^k$ and not at $\hat{\mathbf{x}}^k$.

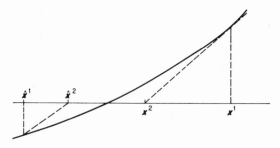

Figure 8.3.2

8.3.5 Assume that the conditions of 8.3.4 hold and that in addition F' satisfies the monotonicity condition $F'(\mathbf{x}) \leq F'(\mathbf{y})$ whenever $\mathbf{x} \leq \mathbf{y}$. Suppose that there is an $\hat{\mathbf{x}}^1$ such that $F\hat{\mathbf{x}}^1 \leq 0$. Then the sequence (7) satisfies

$$\hat{\mathbf{x}}^k \leq \hat{\mathbf{x}}^{k+1} \leq \mathbf{x}^*, \qquad k = 1, 2, \ldots \tag{9}$$

and $\hat{\mathbf{x}}^k \to \mathbf{x}^*$ as $k \to \infty$.

Proof: Since $F'(\mathbf{x}^1)^{-1} \geq 0$ and $F\hat{\mathbf{x}}^1 \leq \mathbf{0}$, we have

$$\hat{\mathbf{x}}^2 = \hat{\mathbf{x}}^1 - F'(\mathbf{x}^1)^{-1}F\hat{\mathbf{x}}^1 \geq \hat{\mathbf{x}}^1.$$

Moreover, by 8.3.3,

$$0 \geq F\hat{\mathbf{x}}^1 = F\hat{\mathbf{x}}^1 - F\mathbf{x}^* \geq F'(\mathbf{x}^*)(\hat{\mathbf{x}}^1 - \mathbf{x}^*)$$

so that multiplication through by $F'(\mathbf{x}^*)^{-1} \geq \mathbf{0}$ yields $\hat{\mathbf{x}}^1 \leq \mathbf{x}^*$. Another application of 8.3.3 shows that

$$F\hat{\mathbf{x}}^1 - F\mathbf{x}^1 \geq F'(\mathbf{x}^1)(\hat{\mathbf{x}}^1 - \mathbf{x}^1)$$

and thus

$$\mathbf{x}^1 \geq \mathbf{x}^1 - F'(\mathbf{x}^1)^{-1}F\mathbf{x}^1 = \hat{\mathbf{x}}^2 + \mathbf{x}^1 - \hat{\mathbf{x}}^1 + F'(\mathbf{x}^1)^{-1}(F\hat{\mathbf{x}}^1 - F\mathbf{x}^1) \geq \hat{\mathbf{x}}^2.$$

Therefore, by assumption $F'(\hat{\mathbf{x}}^2) \leq F'(\mathbf{x}^1)$ and, again using 8.3.3, we have

$$F\hat{\mathbf{x}}^2 \leq F\hat{\mathbf{x}}^1 + F'(\hat{\mathbf{x}}^2)(\hat{\mathbf{x}}^2 - \hat{\mathbf{x}}^1) \leq F\hat{\mathbf{x}}^1 + F'(\mathbf{x}^1)(\hat{\mathbf{x}}^2 - \hat{\mathbf{x}}^1) = 0.$$

By induction, we can prove in a similar way that

$$\hat{\mathbf{x}}^{k-1} \leq \hat{\mathbf{x}}^k \leq \mathbf{x}^*, \qquad F\hat{\mathbf{x}}^k \leq 0, \qquad k = 3, 4, \ldots,$$

and the conclusions of the theorem follow in a manner completely analogous to 8.3.4. $$$

We illustrate the previous two results by the system given by (8.2.7) and (8.2.8). If we assume that g is continuously differentiable and that $g'(s) \geq 0$ for all $s \in (-\infty, \infty)$, then we have shown in 8.1 that $F'(\mathbf{x}) = A + \Phi'(\mathbf{x})$ is an M-matrix and thus, in particular, $F'(\mathbf{x})^{-1} \geq 0$ for all $\mathbf{x} \in R^n$. We now assume, in addition, that g is convex. Then it is easy to see (E8.3.2) that F is also convex. As mentioned in 8.1, it is possible to show that the system $F\mathbf{x} = \mathbf{0}$ has a solution. Hence, all the conditions of 8.3.4 are satisfied.

In order to apply 8.3.5, we note that the convexity of g is sufficient to imply that $F'(\mathbf{x}) \leq F'(\mathbf{y})$ whenever $\mathbf{x} \leq \mathbf{y}$ (E8.3.2), and it remains only to choose the point $\hat{\mathbf{x}}^1$. Take

$$\hat{\mathbf{x}}^1 = \mathbf{x}^1 - A^{-1}F\mathbf{x}^1$$

where, as usual, $\mathbf{x}^1$ is the first Newton iterate, and note that

$$F'(\mathbf{x})A^{-1} = I + \Phi'(\mathbf{x})A^{-1} \geq I$$

since $\Phi'(\mathbf{x}) \geq 0$ and $A^{-1} \geq 0$. Consequently, using **8.3.3** and the fact that we have shown that $F\mathbf{x}^1 \geq 0$, we have

$$F\hat{\mathbf{x}}^1 \leq F\mathbf{x}^1 + F'(\hat{\mathbf{x}}^1)(\hat{\mathbf{x}}^1 - \mathbf{x}^1) = F\mathbf{x}^1 - F(\hat{\mathbf{x}}^1)A^{-1}F\mathbf{x}^1 \leq 0.$$

Finally, we note that if gaussian elimination is being used to solve the linear systems of Newton's method, then the subsidiary sequence (7) can be computed with little additional work.

EXERCISES

E8.3.1   Suppose that $A: R^n \to L(R^n)$ is a matrix-valued function which is continuous at some point $\mathbf{y}$ for which $A(\mathbf{y})$ is nonsingular. Use **2.1.1** to show that $A(\mathbf{x})$ is nonsingular for $\mathbf{x}$ in a neighborhood of $\mathbf{y}$ and that $A(\mathbf{x})^{-1}$ is a continuous function of $\mathbf{x}$ at $\mathbf{y}$.

E8.3.2   Assume that $g: R^1 \to R^1$ is convex. Show that the mappings $\Phi: R^n \to R^n$ and $F: R^n \to R^n$ defined by (8.2.7) and (8.2.8) are convex. Show, moreover, that if g is differentiable then $F'(\mathbf{x}) \leq F'(\mathbf{y})$ whenever $\mathbf{x} \leq \mathbf{y}$.

E8.3.3   Write a computer program to solve the system (8.2.7) by Newton's method. For the choice $g(x) = x^3$, experiment with different starting values for the iteration.†

E8.3.4   Consider the functions $F: R^1 \to R^1$ defined by $Fx = \tan^{-1} x$ and $Fx = 2x + \sin x$. Show that in either case all the conditions of **8.3.4** except the convexity are satisfied, but that the Newton iteration is not globally convergent.

READING

The material of this chapter is taken largely from Ortega and Rheinboldt [1970]. See also Daniel [1971] for the closely related problem of minimization of functionals.

† An open mathematical question is the global convergence of Newton's method for this problem.

PART IV

ROUNDING ERROR

In this last part, we consider the important problem of rounding error, which is, in many ways, the most difficult to analyze of the three basic types of error. Although rounding error is present in essentially every calculation carried out on a computer, we will restrict our attention to the method of gaussian elimination for linear systems of equations and present the famous backward error analysis of J. H. Wilkinson.

CHAPTER 9

ROUNDING ERROR
FOR GAUSSIAN ELIMINATION

9.1 REVIEW OF THE METHOD

In this chapter, we will analyze the effects of rounding error on the most common method for solving a system of linear equations. We begin by reviewing the form and some of the basic variants and properties of this method.

We consider the linear system

$$A\mathbf{x} = \mathbf{b} \tag{1}$$

where $A \in L(R^n)$ is assumed to be nonsingular. Then the "**pure**" **gaussian elimination process** may be described as follows. At the kth stage, we have a "reduced" system

$$A^{(k)}\mathbf{x} = \mathbf{b}^{(k)} \tag{2}$$

where $A^{(1)} = A$ and $\mathbf{b}^{(1)} = \mathbf{b}$, and, in general, $A^{(k)}$ and $\mathbf{b}^{(k)}$ are of the form

$$
A^{(k)} = \begin{bmatrix} a_{11}^{(k)} & \cdots & & \cdots & a_{1n}^{(k)} \\ & \ddots & & & \vdots \\ & & a_{kk}^{(k)} & \cdots & a_{kn}^{(k)} \\ & \bigcirc & \vdots & & \vdots \\ & & a_{nk}^{(k)} & \cdots & a_{nn}^{(k)} \end{bmatrix}, \qquad
\mathbf{b}^{(k)} = \begin{bmatrix} b_1^{(k)} \\ \cdot \\ \cdot \\ \cdot \\ b_n^{(k)} \end{bmatrix}. \tag{3}
$$

In order to obtain the next reduced system we assume that

$$a_{kk}^{(k)} \neq 0 \tag{4}$$

and define the elements of $A^{(k+1)}$ and $\mathbf{b}^{(k+1)}$ by

$$a_{ij}^{(k+1)} = \begin{cases} a_{ij}^{(k)}, & \text{whenever } i \leq k \text{ or } j < k \\ a_{ij}^{(k)} - m_{ik} a_{kj}^{(k)}, & i = k+1, \ldots, n, \quad j = k, \ldots, n \end{cases} \tag{5}$$

and

$$b_i^{(k+1)} = \begin{cases} b_i^{(k)}, & i \leq k \\ b_i^{(k)} - m_{ik} b_k^{(k)}, & i = k+1, \ldots, n \end{cases} \tag{6}$$

where

$$m_{ik} = a_{ik}^{(k)}/a_{kk}^{(k)}, \qquad i = k+1, \ldots, n. \tag{7}$$

The effect of these computations is to introduce zeros into the positions of the kth column below the main diagonal; hence, after $n-1$ stages the resulting matrix $A^{(n)}$ is triangular. This completes the **forward** (or **triangular**) **reduction** phase of the method. To obtain the solution, we then perform the **back substitution**; that is, we solve the triangular system $A^{(n)}\mathbf{x} = \mathbf{b}^{(n)}$ by

$$x_n = b_n^{(n)}/a_{nn}^{(n)} \tag{8}$$

and

$$x_i = (b_i^{(n)} - a_{i,i+1}^{(n)} x_{i+1} - \cdots - a_{i,n}^{(n)} x_n)/a_{ii}^{(n)}, \qquad i = n-1, n-2, \ldots, 1. \tag{9}$$

It is perhaps not immediately obvious that the vector $\mathbf{x}$ obtained in this way is, in fact, the solution of (1), but this will be a consequence of the following matrix representation of the process.

9.1.1 Let m_{ik} be given by (7) and define the vectors

$$\mathbf{m}_k = (0, \ldots, 0, m_{k+1,k}, \ldots, m_{n,k})^{\mathrm{T}}, \qquad k = 1, \ldots, n-1 \tag{10}$$

and matrices

$$M^{(k)} = I - \mathbf{m}_k \mathbf{e}_k^{\mathrm{T}} = \begin{bmatrix} 1 & & & & \\ & 1 & & & \\ & & -m_{k+1,k} & \cdot & \\ & & \vdots & & \\ & & -m_{nk} & & 1 \end{bmatrix} \tag{11}$$

where e_k is the kth coordinate vector. Then the matrices $A^{(k)}$ and vectors $b^{(k)}$ defined above may be represented as

$$A^{(k)} = M^{(k-1)} \cdots M^{(1)}A, \qquad b^{(k)} = M^{(k-1)} \cdots M^{(1)}b, \qquad k = 2, \ldots, n.$$

(12)

Proof: Clearly, (5) shows that

$$A^{(2)} = A - \begin{bmatrix} 0 & \cdots & 0 \\ m_{21}a_{11} & \cdots & m_{21}a_{1n} \\ \vdots & & \vdots \\ m_{n1}a_{11} & \cdots & m_{n1}a_{1n} \end{bmatrix} = M^{(1)}A$$

and similarly for $b^{(2)}$. Continuing this process, we obtain the representation (12). $\$\$\$$

By means of 9.1.1, we see that the original system $Ax = b$ is reduced to the triangular form $A^{(n)}x = b^{(n)}$ by successive multiplications by the matrices $M^{(1)}, \ldots, M^{(n-1)}$. Since each $M^{(k)}$ is obviously nonsingular, the reduced systems (2) all have the same solution.

The product $M^{(n-1)} \cdots M^{(1)}$ is lower triangular, and its diagonal elements are all equal to unity. Therefore

$$L = [M^{(n-1)} \cdots M^{(1)}]^{-1}$$

(13)

is also lower triangular with diagonal elements equal to unity. (See E9.1.3 and also E9.1.4, for an explicit representation of L.) Therefore, since $A^{(n)}$ is upper triangular by construction, 9.1.2 shows that A has been factored into a lower triangular matrix times an upper triangular matrix; that is,

$$A = LU = \begin{bmatrix} 1 & & & \\ l_{21} & 1 & & O \\ \vdots & & \ddots & \\ l_{n1} & \cdots & l_{n,n-1} & 1 \end{bmatrix} \begin{bmatrix} u_{11} & \cdots & u_{1n} \\ & \ddots & \vdots \\ O & & u_{nn} \end{bmatrix}.$$

(14)

Hence, the (pure) gaussian elimination method is mathematically equivalent to the procedure:

(a) Factor A as in (14) (forward reduction)
(b) Solve the triangular system $Ux = L^{-1}b$ (back substitution).

All of the above discussion, including 9.1.1, is, of course, predicated on the assumption (4). We next relate this condition to the matrix A itself. Recall that a **leading principal submatrix** of A is of the form

$$\begin{bmatrix} a_{11} & \cdots & a_{1k} \\ \vdots & & \vdots \\ a_{k1} & \cdots & a_{kk} \end{bmatrix}$$

where k may range from 1 to n.

9.1.2   (Triangular Decomposition Theorem)   Assume that $A \in L(R^n)$ is nonsingular. Then A has a factorization of the form (14) if and only if all leading principal submatrices of A are nonsingular. Moreover, the factorization is unique.

Proof: For the sufficiency, it suffices by our previous discussion to show that the (pure) gaussian elimination process may be carried out and for this it suffices to show that (4) holds. For $k = 1$, this is simply the condition that $a_{11} \neq 0$; that is, the first principal leading submatrix is nonsingular. Now suppose that $a_{ii}^{(i)} \neq 0$, $i = 1, \ldots, k - 1$ so that we have been able to compute $A^{(2)}, \ldots, A^{(k)}$ and the $M^{(1)}, \ldots, M^{(k-1)}$ as given by (11). Write (12) in the form

$$A^{(k)} = \begin{bmatrix} A_{11}^{(k)} & A_{12}^{(k)} \\ A_{21}^{(k)} & A_{22}^{(k)} \end{bmatrix} = \begin{bmatrix} M_{11}^{(k-1)} & 0 \\ M_{21}^{(k-1)} & M_{22}^{(k-1)} \end{bmatrix} \cdots \begin{bmatrix} M_{11}^{(1)} & 0 \\ M_{21}^{(1)} & M_{22}^{(1)} \end{bmatrix} \begin{bmatrix} A_{11} & A_{12} \\ A_{21} & A_{22} \end{bmatrix} \quad (15)$$

where $A_{11}^{(k)}$ is the leading principal submatrix of order k of $A^{(k)}$ and the matrices $M^{(k-1)}, \ldots, M^{(1)}, A$ are partitioned accordingly. Since the $M^{(i)}$ are lower triangular, it follows from (15) that

$$A_{11}^{(k)} = M_{11}^{(k-1)} \cdots M_{11}^{(1)} A_{11}.$$

But all $M_{11}^{(i)}$ are nonsingular and A_{11} is nonsingular by assumption. Hence, $A_{11}^{(k)}$ is also nonsingular and thus $\det A_{11}^{(k)} = a_{11}^{(1)} \cdots a_{kk}^{(k)} \neq 0$, so that (4) holds.

For the necessity, we use a partition similar to (15) and write (14) as

$$\begin{bmatrix} L_{11} & 0 \\ L_{21} & L_{22} \end{bmatrix} \begin{bmatrix} U_{11} & U_{12} \\ 0 & U_{22} \end{bmatrix} = \begin{bmatrix} A_{11} & A_{12} \\ A_{21} & A_{22} \end{bmatrix}$$

where L_{11}, U_{11}, and A_{11} are all $k \times k$. Since A is nonsingular, both L and U, and hence L_{11} and U_{11}, are nonsingular. But then $A_{11} = L_{11} U_{11}$ is nonsingular.

Finally, the uniqueness is proved by assuming that $L_1 U_1 = L_2 U_2 = A$ are two factorizations of the form (14). Then $B = L_1^{-1} L_2 = U_1 U_2^{-1}$ is both upper and lower triangular with diagonal elements equal to unity. That is, $B = I$ so that $L_1 = L_2$ and $U_1 = U_2$. $\$\$\$$

As an immediate corollary of 9.1.1 and 9.1.2 we have:

9.1.3   Assume that $A \in L(R^n)$ is nonsingular. Then the (pure) gaussian elimination process (2)–(7) can be carried out if and only if all leading principal submatrices of A are nonsingular.

The above result shows that pure gaussian elimination can be carried out only if the restrictive principal submatrix condition is satisfied. (See E9.1.5 for two classes of matrices for which this condition is automatically satisfied.) For an arbitrary nonsingular matrix A, a modification of the process must be made and the simplest is the following. If at the kth stage, $a_{kk}^{(k)} = 0$, interchange the kth row of $A^{(k)}$ with a row for which $a_{ik}^{(k)} \neq 0$ for some $i > k$. Clearly, such a row must exist since otherwise $A^{(k)}$ is singular which, by (12), implies that A itself is singular. We shall call this interchange process **mathematical pivoting**. The above discussion is summarized in:

9.1.4   If $A \in L(R^n)$ is nonsingular, then gaussian elimination with mathematical pivoting can always be carried out.

We note that these row interchanges may be viewed in matrix form as a multiplication by a permutation matrix; that is, in place of (12) we now have

$$A^{(k)} = M^{(k-1)} P^{(k-1)} \cdots M^{(1)} P^{(1)} A$$

where the permutation matrix $P^{(i)}$ will be the identity if no interchange is required. Hence

$$A = [M^{(n-1)} P^{(n-1)} \cdots M^{(1)} P^{(1)}]^{-1} A^{(n)}$$

is the composition corresponding to (14). Again $A^{(n)}$ is upper triangular, but the first factor is no longer necessarily lower triangular.

EXERCISES

E9.1.1 Assume that $A \in L(R^n)$ is nonsingular. Verify the **Sherman–Morrison** formula:

$$(A + \mathbf{u}\mathbf{v}^T)^{-1} = A^{-1} - \alpha^{-1} A^{-1} \mathbf{u}\mathbf{v}^T A^{-1}$$

where $\mathbf{u}, \mathbf{v} \in R^n$ are such that $\alpha = 1 + \mathbf{v}^T A^{-1} \mathbf{u} \neq 0$.

E9.1.2 Show that

$$(I - \mathbf{m}_k \mathbf{e}_k^T)^{-1} = I + \mathbf{m}_k \mathbf{e}_k^T$$

where $\mathbf{e}_k$ is the kth coordinate vector and $\mathbf{m}_k$ is given by (10).

E9.1.3 Show that the inverse of a nonsingular lower triangular matrix L is lower triangular. Show, moreover, that if L has diagonal elements equal to unity then so does L^{-1}.

E9.1.4 Let $M^{(1)}, \ldots, M^{(n-1)}$ be given by (10) and (11). Use **E9.1.2** to show that

$$L = [M^{(n-1)} \cdots M^{(1)}]^{-1} = \begin{bmatrix} 1 & & & & \\ m_{21} & 1 & & & \\ \cdot & m_{32} & \ddots & & \\ \cdot & & \vdots & 1 & \\ m_{n1} & m_{n2} & \cdots & m_{nn-1} & 1 \end{bmatrix}.$$

E9.1.5 (a) Assume that $A \in L(R^n)$ is symmetric positive definite. Use **E3.3.3** to show that all leading principal submatrices of A are nonsingular.

(b) Assume that $A \in L(R^n)$ is strictly diagonally dominant. Show that all leading principal submatrices are nonsingular.

Conclude, in either case, that pure gaussian elimination may be carried out.

E9.1.6 The **Cholesky decomposition** of a symmetric positive definite matrix $A \in L(R^n)$ is $A = LL^T$, where L is lower triangular with positive diagonal elements. Use the fact that the leading principal submatrices of A are nonsingular (**E9.1.5**) to show that this decomposition is possible.

E9.1.7 Write out the formulas to apply gaussian elimination to the matrix equation $AX = B$ where A is $n \times n$, and X and B are $n \times m$. Show that this may be done with only one triangular decomposition of A, no matter how large m is. Discuss the special case $m = n$ and $B = I$.

E9.1.8 Let $A \in L(R^n)$ be nonsingular. Show that there is a permutation matrix P such that the leading principal submatrices of PA are all nonsingular.

E9.1.9 Show that gaussian elimination requires approximately $n^3/3$ multiplications or additions for large n.

E9.1.10 Assume that pure gaussian elimination can be carried out. Show that $\det A = a_{11}a_{22}^{(2)} \cdots a_{nn}^{(n)}$. Discuss the necessary modifications to this formula if interchanges are used.

E9.1.11 If A_{11} is nonsingular, show that

$$\begin{bmatrix} A_{11} & A_{12} \\ A_{21} & A_{22} \end{bmatrix} = \begin{bmatrix} I & 0 \\ A_{21}A_{11}^{-1} & I \end{bmatrix} \begin{bmatrix} A_{11} & A_{12} \\ 0 & A_{22} - A_{21}A_{11}^{-1}A_{12} \end{bmatrix}.$$

9.2 ROUNDING ERROR AND INTERCHANGE STRATEGIES

The gaussian elimination method discussed in the previous section produces the exact solution of the linear system $A\mathbf{x} = \mathbf{b}$ with finitely many arithmetical operations. Hence there is no discretization or convergence error in this method, and thus the only possible error is the rounding error in the arithmetic operations. We will discuss in this section certain effects of rounding error and various interchange strategies needed to mitigate these effects. In the next section, we shall give a rigorous rounding error analysis.

We shall assume that we work with a machine in which numbers are represented in the form $r \cdot \beta^s$ where β is the base of the number system (for example, $\beta = 2$ for binary arithmetic, and $\beta = 10$ for decimal arithmetic) and r is the "fraction" or "mantissa" consisting of t digits. We assume that s is allowed to range between $-\infty$ and $+\infty$ so that our machine is idealized in the sense that no underflow or overflow can occur.

For machine numbers x and y, we will denote the machine arithmetic operations by $fl(xy)$, $fl(x/y)$ and $fl(x \pm y)$, and we will assume that

$$fl(xy) = xy(1 + \varepsilon), \qquad |\varepsilon| \leq \beta^{-t+1} \tag{1}$$

$$fl(x/y) = \frac{x}{y}(1 + \varepsilon), \qquad |\varepsilon| \leq \beta^{-t+1} \qquad (2)$$

and

$$fl(x \pm y) = (x \pm y)(1 + \varepsilon), \qquad |\varepsilon| \leq \beta^{-t+1} \qquad (3)$$

which imply the inequalities

$$|fl(xy) - xy| \leq |xy|\beta^{-t+1} \qquad (4)$$

$$|fl(x/y) - x/y| \leq \frac{|x|}{|y|}\,\beta^{-t+1} \qquad (5)$$

and

$$|fl(x \pm y) - (x \pm y)| \leq |x \pm y|\beta^{-t+1}. \qquad (6)$$

The equations (1)–(3) are the formal axioms for the relation of the computed quantities to the exact quantities and reflect a machine in which the exact result is *chopped* (not rounded) to t β-digits. Note that (1)–(3) give relative error bounds. For division and multiplication, this will be realistic on essentially all machines, but for addition and subtraction (3) is usually not valid and must be replaced by

$$fl(x \pm y) = x(1 + \varepsilon_1) \pm y(1 + \varepsilon_2), \qquad |\varepsilon_i| \leq \beta^{-t+1}. \qquad (7)$$

Here, the computed result is equal to the result of exact arithmetic on quantities with small relative errors. Although (7) is usually more realistic, for simplicity we will use the assumption (3).

Before proceeding further, we give some examples of the use of (1).

Example 1   Let $\beta = 10$, $t = 4$, $x = 0.9999$, and $y = 0.1111 \cdot 10^3$. Then

$$xy = 0.11108889 \cdot 10^3, \qquad fl(xy) = 0.1110 \cdot 10^3$$

so that

$$fl(xy) = xy(1 + \varepsilon), \qquad \varepsilon = -0.0008 \cdots.$$

Note that $|\varepsilon| \leq 10^{-3} = \beta^{-t+1}$.

Example 2   Let $\beta = 2$, $t = 4$, $x = 0.1010$ and $y = 0.1111 \cdot 2^3$ where x and y are given in the binary number system. Then

$$xy = 0.10010110 \cdot 2^3, \qquad fl(xy) = 0.1001 \cdot 2^3$$

so that

$$\frac{|xy - fl(xy)|}{|xy|} \doteq 0.11 \cdot 2^{-4} < 2^{-3}$$

where the last quantity is the bound given by (4). Note that in this case the bound is considerably larger than the actual error.

We return now to gaussian elimination and give some simple examples of the effect of rounding error. Consider the system

$$1.00 \cdot 10^{-4}x_1 + 1.00x_2 = 1.00 \tag{8}$$
$$1.00x_1 + 1.00x_2 = 2.00$$

and assume that we carry out gaussian elimination on a three-digit decimal machine; that is, in the above notation, $t = 3$ and $\beta = 10$. From (9.1.5) –(9.1.7) we have

$$fl(m_{21}) = fl(a_{21}/a_{11}) = 10^4 \qquad \text{(exact)}$$
$$fl(a_{22}^{(2)}) = fl(1.00 - 1.00 \cdot 10^4) = -1.00 \cdot 10^4 \qquad \text{(correct to three digits)}$$
$$fl(b_2^{(2)}) = fl(2.00 - 1.00 \cdot 10^4) = -1.00 \cdot 10^4 \qquad \text{(correct to three digits)}.$$

Therefore the computed reduced triangular system is

$$1.00 \cdot 10^{-4}x_1 + 1.00x_2 = 1.00$$
$$-1.00 \cdot 10^4 = -1.00 \cdot 10^4.$$

The back substitution is done exactly on our machine and produces the computed results

$$fl(x_2) = 1.00, \qquad fl(x_1) = 0.$$

But the exact solution of (8) is

$$x_2 = 0.99990 \cdots, \qquad x_1 = 1.00010 \cdots \tag{9}$$

so that while $fl(x_2)$ is a very good approximation to x_2, $fl(x_1)$ is very poor.

It is of interest to view the above results in the context of Chapter 2. First note that the coefficient matrix of (8) is very well conditioned (E9.2.1). However, in our three-digit machine $fl(x_1) = 0$ and $fl(x_2) = 1.00$ would be the computed solution of *any* system of the form

$$1.00 \cdot 10^{-4}x_1 + 1.00x_2 = 1.00$$
$$1.00x_1 + a_{22}x_2 = b_2$$

where $|a_{22}| < 100$. and $|b_2| < 100$. Clearly, such large perturbations of even a well-conditioned system must be expected to produce large changes in the solution.

Consider now the simple artifice of interchanging the equations of (8); that is, consider the system

$$1.00x_1 + 1.00x_2 = 2.00$$
$$1.00 \cdot 10^{-4}x_1 + 1.00x_2 = 1.00.$$

Then gaussian elimination produces

$$fl(m_{21}) = 10^{-4} \quad \text{(exact)}$$

$$fl(a_{22}^{(2)}) = fl(1.00 - 1.00 \cdot 10^{-4}) = 1.00 \quad \text{(correct to three digits)}$$

$$fl(b_2^{(2)}) = fl(1.00 - 2.00 \cdot 10^{-4}) = 1.00 \quad \text{(correct to three digits)}$$

so that

$$fl(x_2) = 1.00, \qquad fl(x_1) = 1.00$$

which are excellent approximations to the exact solutions.

This simple example leads us to the following general interchange strategy.

9.2.1 (Partial Pivoting Strategy)   At the kth stage of gaussian elimination in which the subsystem

$$a_{kk}^{(k)}x_k + \cdots + a_{kn}^{(k)}x_n = b_k^{(k)}$$
$$\vdots \qquad\qquad\qquad (10)$$
$$a_{nk}^{(k)}x_k + \cdots + a_{nn}^{(k)}x_n = b_n^{(k)}$$

is still to be reduced, let i be any index between k and n for which

$$|a_{ik}^{(k)}| = \max_{k \leq j \leq n} |a_{jk}^{(k)}|$$

and interchange the ith and kth rows.

Note that when the partial pivoting strategy is used, all of the multipliers are less than or equal to one.

Even though gaussian elimination with the partial pivoting strategy has proven to be quite reliable in practice, it is by no means foolproof. We first note a danger which is not really the fault of the interchange strategy. Consider the system

$$1.00 \cdot 10^{-4}x_1 + 1.00x_2 = 1.00$$
$$1.00 \cdot 10^{-4}x_1 + 1.00 \cdot 10^{-4}x_2 = 2.00 \cdot 10^{-4}. \tag{11}$$

Since the elements of the first column are equal, no interchange is necessary under the partial pivoting strategy and if one uses our hypothetical three-digit machine, it is easy to see (E9.2.2) that $fl(x_1) = 0$ and $fl(x_2) = 1.00$. But (11) is precisely the system (8) with the second equation scaled by 10^{-4}, and hence has the exact solution (9). Therefore, we have obtained the same erroneous solution as we did with gaussian elimination with no interchanges applied to the system (8). Note, however, that the coefficient matrix of (11) is now very badly conditioned (E9.2.1).

The above example underlines the importance of proper scaling of the coefficient matrix. This can take different forms, some of which are given in the following.

9.2.2 Definition   The matrix $A \in L(R^n)$ is **row-equilibrated** if the maximum elements of each row are equal in modulus; that is, if

$$\max_{1 \le j \le n} |a_{1j}| = \max_{1 \le j \le n} |a_{2j}| = \cdots = \max_{1 \le j \le n} |a_{nj}|. \tag{12}$$

It is **column-equilibrated** if

$$\max_{1 \le j \le n} |a_{j1}| = \max_{1 \le j \le n} |a_{j2}| = \cdots = \max_{1 \le j \le n} |a_{jn}| \tag{13}$$

and **equilibrated** if it is both row and column equilibrated.

Given an arbitrary $n \times n$ matrix A it is essentially trivial to row equilibrate it or column equilibrate it. In practice, it is sufficient that the equalities in (12) or (13) hold only approximately and the scaling should be done by powers of the number base β so that no rounding error is introduced by these preliminary operations. The important thing is to avoid badly scaled problems such as (11).

We next show that the partial pivoting strategy can fail badly even on an equilibrated matrix. For odd n, define the $n \times n$ matrix†

---

† Due to J. Wilkinson, Error Analysis of Direct Methods of Matrix Inversion, *J. Assoc. Comp. Mach.* **8** (1961), 281–330.

$$A = \begin{bmatrix} 1 & 0 & 0 & \cdot & \cdot & \cdot & 0 & 1 \\ 1 & 1 & 0 & \cdot & \cdot & \cdot & 0 & -1 \\ -1 & 1 & 1 & \cdot & & & & 1 \\ 1 & -1 & & \cdot & \cdot & & & \cdot \\ -1 & 1 & & & \cdot & \cdot & 0 & \cdot \\ \vdots & \vdots & & & & \cdot & & (-1)^n \\ (-1)^n & (-1)^{n-1} & & \cdot & \cdot & \cdot & 1 & 1 \end{bmatrix} \tag{14}$$

and consider the linear system $A\mathbf{x} = \mathbf{e}_1$, whose solution is the first column of A^{-1}. It is easy to verify (E9.2.4) that the exact solution is

$$\mathbf{x} = \tfrac{1}{2}(1, 0, \ldots, 0, 1)^{\mathrm{T}}. \tag{15}$$

Now define the matrix

$$B = A + \tfrac{1}{2}\mathbf{e}_n \mathbf{e}_n^{\mathrm{T}} \tag{16}$$

that is, B is identical with A except that the (n, n) element is $\tfrac{3}{2}$ in place of 1. The solution of the system $B\mathbf{y} = \mathbf{e}_1$ is (E9.2.5)

$$\mathbf{y} = (2^{-1}, 0, \ldots, 0, 2^{-1})^{\mathrm{T}}$$
$$- (-2^{-n-1}, 2^{-n}, -2^{-n+1}, \ldots, 2^{-3}, 2^{-n-1})^{\mathrm{T}}/(1 + 2^{-n}) \tag{17}$$

so that the components of $\mathbf{x}$ and $\mathbf{y}$ differ by at most 2^{-3}.

Now suppose that gaussian elimination with the partial pivoting strategy is applied to the two systems $A\mathbf{x} = \mathbf{e}_1$ and $B\mathbf{y} = \mathbf{e}_1$. It is easy to see that no interchanges are needed and, because of the structure of A, the elements in the reduced matrices remain the same except in the column whose elements are being set to zero, and the last column. The elements in the last column successively build up during the forward reduction and, in particular, the (n, n) element becomes 2^{n-1} for A and $2^{n-1} + \tfrac{1}{2}$ for B. Clearly, the reduced triangular matrices for A and B differ in only this one element. The elements in the right-hand sides of the equations also build up, and, for both systems, the final right-hand side is given by

$$(1, -1, 2, -2^2, \ldots, -2^{n-3}, 2^{n-2}).$$

Assume that the above computations have been done on a binary machine with an n digit fraction, for example, $n = 27$. Then the (n, n) element of the reduced system for $B\mathbf{y} = \mathbf{e}_1$ must be truncated from $2^{n-1} + \tfrac{1}{2}$ to 2^{n-1}, while all other elements of the reduced systems will be exact.

Hence the reduced triangular systems which are computed in this machine will be identical and therefore the computed solutions of $Ax = e_1$ and $By = e_1$ will also be identical. It is easy to see (E9.2.6) that no rounding errors are committed during the computation on $Ax = e_1$ and, hence, both computed solutions are given by (15). Therefore, the errors in the computed elements of the solution of $By = e_1$ are almost as much as 2^{-3}, or 25% of $\|x\|_\infty$.

The difficulty encountered in the previous example may be circumvented by the following more powerful interchange strategy.

9.2.3 (Complete Pivoting Strategy) At the kth stage of gaussian elimination in which the subsystem given by (10) is still to be reduced, let i and j be any indices between k and n for which

$$|a_{ij}^{(k)}| = \max_{k \le l,\, m \le n} |a_{lm}^{(k)}|$$

and interchange the ith and kth rows of the subsystem and then interchange the jth and kth columns.

The effect of the complete pivoting strategy is that, after the interchanges, the largest element in the entire submatrix still to be reduced is in the (k, k) position. We shall see in the next section that much better error bounds for the rounding error may be given for this interchange strategy than for partial pivoting.

EXERCISES

E9.2.1 Compute the l_1 and l_2 condition numbers of the coefficient matrices of the systems (8) and (11).

E9.2.2 Use gaussian elimination on the three-digit machine of the text for the system (11) and show that $fl(x_1) = 0$, $fl(x_2) = 1.00$.

E9.2.3 Row equilibrate the coefficient matrix of (11).

E9.2.4 Show, by direct computation, that the solution of $Ax = e_1$, where A is given by (14), is $x = \frac{1}{2}(1, 0, \ldots, 0, 1)^T$.

E9.2.5 Let A be given by (14). Show that the last column of A^{-1}, that is, the solution of $Ax = e_n$, is given by

$$x = (-2^{-n+1}, 2^{-n+2}, \ldots, -2^{-2}, 2^{-1}, 2^{-n+1}).$$

Use this in the Sherman–Morrison formula of E9.1.1 to show that the solution of $B\mathbf{y} = \mathbf{e}_1$, where B is as in (16), is given by (17).

E9.2.6 Let A be given by (14) and assume that gaussian elimination with partial pivoting is carried out for the system $A\mathbf{x} = \mathbf{e}_1$ on a machine with an n digit fraction. Show that no rounding error is committed during the computation and, hence, the computed solution is the exact solution (15). Verify this on a computer.

E9.2.7 Write a computer program to carry out gaussian elimination with either the partial pivoting or complete pivoting strategies. Verify that a good solution to $B\mathbf{x} = \mathbf{e}_1$, where B is given by (16), can be obtained using complete pivoting.

9.3 BACKWARD ERROR ANALYSIS

There are two general approaches to rounding error analysis. In the first, called **forward error analysis**, the cumulative rounding error is bounded according to the sequence of arithmetic operations to be carried out. For example, consider the error in the product abc under the assumption that the basic inequality (9.2.4) holds. Then $|fl(ab) - ab| \le |ab|\beta^{-t+1}$ so that

$$
\begin{aligned}
|fl(fl(ab)c) - abc| &\le |fl(fl(ab)c) - fl(ab)c| + |fl(ab)c - abc| \\
&\le |fl(ab)c|\beta^{-t+1} + |abc|\beta^{-t+1} \\
&\le |abc|(1 + \beta^{-t+1})\beta^{-t+1} + |abc|\beta^{-t+1} \\
&\le |abc|(2 + \beta^{-t+1})\beta^{-t+1}.
\end{aligned}
$$

Clearly, the analysis of a lengthy calculation in this manner would be very tedious.

The second approach, **backward error analysis**, attempts to show that the computed solution is the result of exact computation on different data. This is the approach we shall follow in this section and, in particular, we will show that the effect of rounding error in solving the linear system $A\mathbf{x} = \mathbf{b}$ by gaussian elimination is that the computed solution $\hat{\mathbf{x}}$ is the *exact solution* of a system of the form

$$(A + F)\hat{\mathbf{x}} = \mathbf{b}. \tag{1}$$

An estimate for the actual error, $\hat{\mathbf{x}} - A^{-1}\mathbf{b}$, may then be obtained by the results of Chapter 2 in terms of $\|F\|$. Surprisingly, this analysis may be carried out rather simply regardless of the size of the matrix.

In the sequel, we shall assume that the rounding error axioms (9.2.1)–(9.2.3) hold as well as the inequalities (9.2.4)–(9.2.6) which they imply. We will also assume that the matrices

$$A^{(k)} = \begin{bmatrix} a_{11}^{(k)} & & & \cdots & & a_{1n}^{(k)} \\ & \ddots & & & & \vdots \\ & & a_{kk}^{(k)} & \cdots & a_{kn}^{(k)} \\ & & \vdots & & \vdots \\ & & a_{nk}^{(k)} & \cdots & a_{nn}^{(k)} \end{bmatrix} \tag{2}$$

discussed in Section 9.1, are those *which are actually produced by the machine*.

Consider now the first stage in which $A^{(2)}$ is computed from $A^{(1)} = A$. If we denote the computed multipliers of (9.1.7) by $\bar{m}_{i1}$, we have

$$\bar{m}_{i1} = fl(a_{i1}/a_{11}) = \frac{a_{i1}}{a_{11}}(1 + \eta_{i_1}), \qquad |\eta_{i_1}| \le \beta^{-t+1}$$

and the computed elements $a_{ij}^{(2)}$ are then given by

$$a_{ij}^{(2)} = fl(a_{ij} - fl(\bar{m}_{i1}a_{1j})) = [a_{ij} - fl(\bar{m}_{i1}a_{1j})](1 + \alpha_{ij}^{(1)})$$
$$= [a_{ij} - \bar{m}_{i1}a_{1j}(1 + \gamma_{ij}^{(1)})](1 + \alpha_{ij}^{(1)}), \qquad i, j = 2, \ldots, n \tag{3}$$

where $|\alpha_{ij}^{(1)}| \le \beta^{-t+1}$ and $|\gamma_{ij}^{(1)}| \le \beta^{-t+1}$. By collecting terms in (3), we obtain

$$a_{ij}^{(2)} = a_{ij} - \bar{m}_{i1}a_{1j} + e_{ij}^{(1)}, \qquad i, j = 2, \ldots, n \tag{4}$$

where

$$e_{ij}^{(1)} = \frac{\alpha_{ij}^{(1)}}{1 + \alpha_{ij}^{(1)}} a_{ij}^{(2)} - \bar{m}_{i1}a_{1j}\gamma_{ij}^{(1)}, \qquad i, j = 2, \ldots, n. \tag{5}$$

If we also set

$$e_{i1}^{(1)} = a_{i1}\eta_{i1}, \qquad i = 2, \ldots, n \tag{6}$$

then we may conclude that $A^{(2)}$ is the result of exact arithmetic on the matrix $A + E^{(1)}$ where the first row of $E^{(1)}$ is zero and the other elements of $E^{(1)}$ are given by (5) and (6).

Since $A^{(k+1)}$ is the computed reduced matrix, in exactly the same fashion we can say that $A^{(k+1)}$ is the result of exact calculation on the matrix $A^{(k)} + E^{(k)}$ where

$$E^{(k)} = \begin{bmatrix} 0 & & \cdots & & 0 \\ \vdots & & & & \vdots \\ 0 & & \cdots & & 0 \\ 0 & e^{(k)}_{k+1,k} & \cdots & & e^{(k)}_{k+1,n} \\ \vdots & \vdots & & & \vdots \\ 0 & e^{(k)}_{n,k} & \cdots & & e^{(k)}_{n,n} \end{bmatrix} \tag{7}$$

with

$$e^{(k)}_{i,k} = a^{(k)}_{i,k} \eta_{ik}, \qquad i = k+1,\ldots,n, \quad |\eta_{ik}| \leqslant \beta^{-t+1} \tag{8}$$

$$e^{(k)}_{i,j} = \frac{\alpha^{(k)}_{ij}}{1 + \alpha^{(k)}_{ij}} a^{(k+1)}_{ij} - \bar{m}_{ik} a^{(k)}_{kj} \gamma^{(k)}_{ij}, \qquad i,j = k+1,\ldots,n \tag{9}$$

and

$$|\alpha^{(k)}_{ij}| \leq \beta^{-t+1}, \qquad |\gamma^{(k)}_{ij}| \leq \beta^{-t+1}. \tag{10}$$

Now consider the complete reduction starting with the matrix

$$A + E^{(1)} + \cdots + E^{(n-1)}. \tag{11}$$

Exact arithmetic at the first stage produces the matrix

$$A^{(2)} + E^{(2)} + \cdots + E^{(n-1)} \tag{12}$$

since the matrices $E^{(2)}, \ldots, E^{(n-1)}$ enter the calculation additively (E9.3.1). Similarly, exact arithmetic at the second stage produces

$$A^{(3)} + E^{(3)} + \cdots + E^{(n-1)}$$

and so on until $A^{(n)}$ is the final result of exact arithmetic starting from (11). We summarize our results so far as:

9.3.1   Under the assumption that the elements $a^{(i)}_{ii}$, $i = 1, \ldots, n-1$, are all nonzero, the triangular matrix $A^{(n)}$ produced by machine computation is the result of exact arithmetic applied to the matrix $A + E$ where

$$E = E^{(1)} + \cdots + E^{(n-1)} \tag{13}$$

with the $E^{(i)}$ given by (7)–(10).

Another way of viewing 9.3.1, in the context of the triangular decomposition (9.1.14), is that, in exact arithmetic,

$$A + E = LU \tag{14}$$

where $U = A^{(n)}$ and L is given (E9.1.4) in terms of the computed multipliers $\bar{m}_{ik}$ by

$$L = \begin{bmatrix} 1 & & & & \text{\Large O} \\ \bar{m}_{21} & \cdot & & & \\ \vdots & \ddots & \cdot & & \\ \bar{m}_{n1} & \cdots & \bar{m}_{n,\,n-1} & 1 \end{bmatrix} \tag{15}$$

We shall return to the important question of the magnitude of the elements of E directly. First, let us complete the gaussian elimination process by carrying out the back substitution. This is equivalent to solving the triangular system

$$U\mathbf{x} = \mathbf{y} \tag{16}$$

where $U = A^{(n)}$ and $\mathbf{y}$ is the computed reduced right-hand side of the original system. By proceeding as above, it may be shown† that the computed solution, $\hat{\mathbf{x}}$, of (16) satisfies exactly

$$(U + \delta U)\hat{\mathbf{x}} = \mathbf{y} \tag{17}$$

where δU is a matrix whose elements are bounded according to

$$|\delta U| \le c\beta^{-t+1} \begin{bmatrix} 2|u_{11}| & 2|u_{12}| & 3|u_{13}| & \cdots & n|u_{1n}| \\ & \cdot & \cdot & & \vdots \\ \text{\Large O} & & \cdot & \cdot & 2|u_{n-1,n}| \\ & & & \cdot & 2|u_{nn}| \end{bmatrix}. \tag{18}$$

Here c is a constant‡ only slightly larger than 1, depending on the magnitude of β^{-t}.

The error in the right-hand side $\mathbf{y}$ of (16) may be analyzed in a similar way. In fact, by the decomposition (14) it follows that if the arithmetic in processing the original right-hand $\mathbf{b}$ were done exactly, with the exception of using the computed multipliers $\bar{m}_{ij}$, then the reduced triangular system would be $U\mathbf{x} = L^{-1}\mathbf{b}$. Therefore, $\mathbf{y}$ is the computed solution of

---

† The full analysis is given in Wilkinson [1963].
‡ For example, if we assume that $n\beta^{-t} \le 0.01$ we may take $c = 1.02$.

the system $Lz = b$ and the same analysis as for the system (16) applies to this triangular system. Hence we can say that y is the exact solution of

$$(L + \delta L)y = b \qquad (19)$$

where

$$|\delta L| \leq c\beta^{-t+1}\begin{bmatrix} 2 & & & \\ 2|\overline{m}_{21}| & 2 & & \\ \vdots & & \ddots & \\ n|\overline{m}_{n1}| & \cdots & 2|\overline{m}_{n,n-1}| & 2 \end{bmatrix}. \qquad (20)$$

We can now combine (17) and (19) to conclude that the final computed solution $\hat{x}$ satisfies, in exact arithmetic,

$$(U + \delta U)\hat{x} = (L + \delta L)^{-1}b$$

or

$$(L + \delta L)(U + \delta U)\hat{x} = b.$$

If we expand the left-hand side of this expression and use (14), we see that $\hat{x}$ is the exact solution of

$$(A + F)\hat{x} = b \qquad (21)$$

where

$$F = E + \delta L U + L\,\delta U + \delta L\,\delta U. \qquad (22)$$

We summarize these results as follows.

9.3.2 Assume that the elements $a_{ii}^{(i)}$, $i = 1, \ldots, n-1$ are nonzero. Then the computed solution $\hat{x}$ of $Ax = b$, produced by gaussian elimination, satisfies in exact arithmetic the equation $(A + F)\hat{x} = b$, where F is given by (22) with E defined by (13) and (7)–(10), L is defined by (15) in terms of the computed multipliers, U is the computed upper triangular matrix $A^{(n)}$ and δU and δL are bounded by (18) and (20).

In order to obtain final bounds for the difference $\hat{x} - x$ between the computed and the exact solution, we can apply the error estimate theorem 2.1.2. Thus, if $\|A^{-1}\|\,\|F\| < 1$, we obtain for the relative error

$$\frac{\|x - \hat{x}\|}{\|x\|} \leq \frac{\|A^{-1}\|\,\|F\|}{1 - \|A^{-1}\|\,\|F\|}. \qquad (23)$$

The crucial factor in this estimate which is determined by the computational process is $\|F\|$, and we now investigate this in more detail.

The magnitude of the elements of F depends primarily on the magnitude of the multipliers $\bar{m}_{ij}$ and of the elements $a_{ij}^{(k)}$. Without interchanges of any kind, the multipliers may be arbitrarily large and no useful estimates may be obtained in general. (Recall that an example showing the effects of large multipliers was given in the previous section.) Hence, in the sequel we will assume that either the partial pivoting strategy **9.2.1** or the complete pivoting strategy **9.2.3** is to be used. In either case, the computed multipliers satisfy

$$|\bar{m}_{ik}| \leq 1, \qquad k = 1, \ldots, n - 1, \quad i = k + 1, \ldots, n. \tag{24}$$

It is important to note that interchanges of either rows or columns do not affect the prior analysis in an essential way (of course, as noted in Section 9.1, the matrix L of the decomposition (14) is no longer necessarily triangular). In fact, we shall assume for the purpose of this analysis that all interchanges have been made prior to starting the computation although, of course, we are not able to do this in actual practice.

We next obtain a bound for the matrix E under the assumption (24). We define

$$a = \max_{1 \leq i, j \leq n} |a_{ij}|, \qquad g = \frac{1}{a} \max_{i, j, k} |a_{ij}^{(k)}|. \tag{25}$$

Then, from (8), (9), and (10),

$$|e_{ik}^{(k)}| \leq ag\beta^{-t+1}, \qquad k = 1, \ldots, n - 1, \quad i = k + 1, \ldots, n. \tag{26}$$

and, for $i, j = k + 1, \ldots, n$,

$$|e_{ij}^{(k)}| \leq \frac{\beta^{-t+1}}{1 - \beta^{-t+1}} |a_{ij}^{(k+1)}| + \beta^{-t+1} |a_{ij}^{(k)}| \leq \frac{2}{1 - \beta^{-t+1}} ag\beta^{-t+1}. \tag{27}$$

If we use these estimates in conjunction with (13) and (7) we have

$$|E| \leq ag\gamma \left\{ \begin{bmatrix} 0 & \cdot & \cdot & \cdot & 0 \\ 1 & 2 & \cdots & & 2 \\ \vdots & \vdots & & & \vdots \\ 1 & 2 & \cdots & & 2 \end{bmatrix} + \begin{bmatrix} 0 & \cdot & \cdot & \cdot & 0 \\ 0 & \cdot & \cdot & \cdot & 0 \\ 0 & 1 & 2 & \cdots & 2 \\ \vdots & \vdots & & & \vdots \\ 0 & 1 & 2 & \cdots & 2 \end{bmatrix} + \cdots \right.$$

$$\left. + \begin{bmatrix} 0 & \cdot & \cdot & \cdot & 0 \\ \vdots & & & & \vdots \\ 0 & \cdot & \cdot & \cdot & 0 \\ 0 & \cdots & 0 & 1 & 2 \end{bmatrix} \right\}$$

$$\leq ag\gamma \begin{bmatrix} 0 & \cdot & \cdot & \cdot & 0 \\ 1 & 2 & \cdot & \cdot & \cdot & 2 \\ \cdot & 3 & 4 & \cdots & 4 \\ \vdots & \vdots & & \ddots & \vdots \\ 1 & 3 & \cdot & \cdot & \cdot & 2(n-1) \end{bmatrix},$$

where we have set $\gamma = \beta^{-t+1}/(1 - \beta^{-t+1})$. Therefore,

$$\|E\|_\infty \leq ag\gamma[1 + 3 + \cdots + 2n - 3 + 2(n-1)] \leq n^2 ag \frac{\beta^{-t+1}}{1 - \beta^{-t+1}}. \quad (28)$$

We next obtain bounds for the matrix F of (22). Since $U = A^{(n)}$, we have $|u_{ij}| \leq ag$, $i, j = 1, \ldots, n$, so that

$$\|U\|_\infty \leq nag \quad (29)$$

and, from (18),

$$\|\delta U\|_\infty \leq cag\beta^{-t+1}(2 + 2 + 3 + \cdots + n) \leq \tfrac{1}{2}cag(n+1)^2\beta^{-t+1}. \quad (30)$$

Similarly, (15), (20), and (24) imply that

$$\|L\|_\infty \leq n, \qquad \|\delta L\|_\infty \leq \tfrac{1}{2}c(n+1)^2\beta^{-t+1}. \quad (31)$$

By combining (28)–(31) with (22) we then obtain

$$\|F\|_\infty \leq \|E\|_\infty + \|\delta L\|_\infty \|U\|_\infty + \|L\|_\infty \|\delta U\|_\infty + \|\delta L\|_\infty \|\delta U\|_\infty$$

$$\leq n^2 ag \frac{\beta^{-t+1}}{1 - \beta^{-t+1}} + cn(n+1)^2 ag\beta^{-t+1} + \frac{c^2(n+1)^4}{4} ag(\beta^{-t+1})^2. \quad (32)$$

In order for this bound to be at all useful we must have the wordlength t sufficiently large relative to n, and we shall assume that $n^2\beta^{-t} \ll 1$. Then $(1 - \beta^{-t+1})^{-1} \leq 2$, and since $c \doteq 1$ and $a \leq \|A\|_\infty$, the estimate (32) may be approximated by

$$\|F\|_\infty \lesssim 2(n+1)^3\|A\|_\infty g\beta^{-t} \quad (33)$$

which shows the basic factors on which $\|F\|_\infty$ depends. The quantity $\|A\|_\infty$ is simply a normalizing factor and β^{-t} reflects, of course, the wordlength of the machine. The crucial factors, then, are n^3 and g. The quantity g, defined by (25), may be interpreted as a **growth factor** of the elements of the successive reduced matrices $A^{(k)}$. For the partial pivoting strategy, it is easy to give a bound for g. In fact, since the multipliers satisfy $|m_{ij}| \leq 1$, we have

$$\max_{i,j} |a_{ij}^{(k+1)}| = \max_{i,j} |a_{ij}^{(k)} - m_{ik} a_{kj}^{(k)}| \leq 2 \max_{i,j} |a_{ij}^{(k)}|$$

and it follows that

$$g \leq 2^{n-1} \qquad \text{(partial pivoting).} \qquad (34)$$

This estimate also holds for the complete pivoting strategy but a much better estimate, which we state without proof,† may be given:

$$g \leq n^{1/2}(2 \cdot 3^{1/2} 4^{1/3} \cdots n^{1/(n-1)})^{1/2} \qquad \text{(complete pivoting).} \qquad (35)$$

There is a vast difference between the estimates (34) and (35) when n is large. For example, if $n = 100$

$$2^{n-1} \doteq 10^{30}, \qquad n^{1/2}(2 \cdots n^{1/(n-1)})^{1/2} \doteq 3300.$$

Moreover, the estimate (34) is sharp in that for certain matrices it is possible that $g = 2^{n-1}$; an example of such a matrix is given by (9.2.14).

EXERCISES

E9.3.1   Verify that exact arithmetic in the first stage of gaussian elimination produces the matrix (12) from (11).

E9.3.2   Show that equality holds in (34) for the matrix of (9.2.14).

9.4   ITERATIVE REFINEMENT

Let $\mathbf{x}^0$ be the approximate solution of the linear system $A\mathbf{x} = \mathbf{b}$ which has been computed by gaussian elimination. It is sometimes possible to obtain a better approximation by means of the following iterative procedure, usually called **iterative refinement**: Given the kth iterate $\mathbf{x}^k$:

(a)   Compute the residual vector $\mathbf{r}^k = \mathbf{b} - A\mathbf{x}^k$ in double precision arithmetic
(b)   Solve the system $A\mathbf{y}^k = \mathbf{r}^k$ and set $\mathbf{x}^{k+1} = \mathbf{x}^k + \mathbf{y}^k$.

† See Wilkinson, Error Analysis of Direct Methods of Matrix Inversion, *J. Assoc. Comp. Mach.* **8** (1961), 281–330. A long-standing conjecture of Wilkinson is that (35) may be improved to $g(n) \leq n$.

Since the coefficient matrix A is always the same in all of the systems $A\mathbf{x} = \mathbf{b}$, $A\mathbf{y}^0 = \mathbf{r}^0$, ..., we need only do the forward reduction of A once provided the multipliers and interchange information are saved. The successive right-hand sides $\mathbf{b}$, $\mathbf{r}^0$, $\mathbf{r}^1$, ... are then reduced using these multipliers.

We next write this iterative process in a functional form. We showed in the previous section that the computed solution $\hat{\mathbf{z}}$ of the system $A\mathbf{z} = \mathbf{c}$ satisfies $(A + F)\hat{\mathbf{z}} = \mathbf{c}$ where F is a matrix which is composed of two parts: the matrix E which depends only on A, and the matrix $F - E$ which results from the reduction of the right-hand side of the system and from the back substitution. This latter matrix, and hence F also, depends upon the solution of the system, and, therefore, the vectors $\mathbf{y}^k$ above satisfy, in exact arithmetic, equations of the form

$$(A + F_k)\mathbf{y}^k = \mathbf{r}^k \tag{1}$$

where the subscript on F denotes its dependence on $\mathbf{y}^k$. Hence, the vectors $\mathbf{x}^k$ satisfy

$$\mathbf{x}^{k+1} = \mathbf{x}^k + (A + F_k)^{-1}\mathbf{r}^k, \qquad k = 0, 1, \ldots \tag{2}$$

assuming, of course, that the inverses exist.

In the following analysis, we shall assume *that $\mathbf{r}^k$ is evaluated exactly*; since, in practice, we insist that the $\mathbf{r}^k$ be computed in double precision arithmetic, this is not an unreasonable assumption. Therefore,

$$\mathbf{r}^k = \mathbf{b} - A\mathbf{x}^k$$

and (2) becomes

$$\mathbf{x}^{k+1} = \mathbf{x}^k + (A + F_k)^{-1}(\mathbf{b} - A\mathbf{x}^k) = (A + F_k)^{-1}(F_k\mathbf{x}^k + \mathbf{b}). \tag{3}$$

Now observe that if $F_k = 0$, then $\mathbf{x}^{k+1} = A^{-1}\mathbf{b}$; that is, if the system $A\mathbf{y}^k = \mathbf{r}^k$ could be solved exactly, $\mathbf{x}^k + \mathbf{y}^k$ would be the exact solution of $A\mathbf{x} = \mathbf{b}$. Note, also, that if $\mathbf{x}^k = A^{-1}\mathbf{b}$, then

$$\mathbf{x}^{k+1} = (A + F_k)^{-1}(F_k A^{-1}\mathbf{b} + \mathbf{b}) = A^{-1}\mathbf{b}. \tag{4}$$

This latter property shows that the iteration (2) is consistent in the sense of the following definition.

9.4.1 Definition Let $G_k : R^n \to R^n$ be a sequence of mappings. Then the iterative method

$$\mathbf{x}^{k+1} = G_k\mathbf{x}^k, \qquad k = 0, 1, \ldots$$

is **consistent** with x^* if

$$\mathbf{x}^* = G_k \mathbf{x}^*, \qquad k = 0, 1, \ldots.$$

Suppose that in the iteration (3) the matrices F_k were independent of k, that is, $F_k = F$, $k = 0, 1, \ldots$. Then Theorem 7.1.1 would show that the iterates $\mathbf{x}^k$ converge to $\mathbf{x}^* = A^{-1}\mathbf{b}$ if $\rho\{(A + F)^{-1}F\} < 1$. This theorem does not extend directly to iterative methods of the form

$$\mathbf{x}^{k+1} = H_k \mathbf{x}^k + \mathbf{d}^k, \qquad k = 0, 1, \ldots, \tag{5}$$

with variable iteration matrices H_k (see **E9.4.1**), but the following simple result will suffice for our purposes.

9.4.2   Let $\{H_k\} \subset L(R^n)$ be a sequence of matrices such that

$$\|H_k\| \leq \alpha < 1, \qquad k = 0, 1, \ldots \tag{6}$$

and suppose that the iteration (5) is consistent with $\mathbf{x}^*$. Then $\mathbf{x}^k \to \mathbf{x}^*$ as $k \to \infty$ for any $\mathbf{x}^0$.

Proof: By consistency, it follows that

$$\mathbf{x}^{k+1} - \mathbf{x}^* = H_k(\mathbf{x}^k - \mathbf{x}^*), \qquad k = 0, 1, \ldots$$

so that

$$\|\mathbf{x}^{k+1} - \mathbf{x}^*\| \leq \alpha\|\mathbf{x}^k - \mathbf{x}^*\| \leq \cdots \leq \alpha^{k+1}\|\mathbf{x}^0 - \mathbf{x}^*\|. \quad \$\$\$$$

We now apply this result to the iterative process (3).

9.4.3   Let $A \in L(R^n)$ be nonsingular and suppose there is a constant γ for which

$$\|F_k\| \leq \gamma < \frac{1}{2\|A^{-1}\|}, \qquad k = 0, 1, \ldots. \tag{7}$$

Then the matrices $A + F_k$, $k = 0, 1, \ldots$ are all nonsingular and the iterates (3) converge to $\mathbf{x}^* = A^{-1}\mathbf{b}$ for any $\mathbf{x}^0$.

Proof: By the perturbation lemma 2.1.1, we have that $A + F_k$ is nonsingular and that

$$\|(A + F_k)^{-1}\| \leq \frac{\|A^{-1}\|}{1 - \|A^{-1}\|\,\|F_k\|} \leq \frac{\|A^{-1}\|}{1 - \gamma\|A^{-1}\|}, \qquad k = 0, 1, \ldots.$$

Therefore, if we set $H_k = (A + F_k)^{-1}F_k$ and $\beta = \gamma\|A^{-1}\|$ we obtain

$$\|H_k\| \le \frac{\beta}{1-\beta} < 1, \qquad k = 0, 1, \ldots .$$

Hence, 9.4.2 applies. $\$\$\$$

In the previous section we obtained the approximate estimate

$$\|F_k\|_\infty \lesssim 2(n + 1)^3\|A\|_\infty g\beta^{-t} \tag{8}$$

under the assumption that $n\beta^{-t} \ll 1$. Therefore, in order that (7) hold we need

$$2(n + 1)^3 K_\infty(A)g\beta^{-t} < \tfrac{1}{2}$$

where $K_\infty(A)$ is the condition number of A. This inequality states, roughly, that the iterative refinement procedure will converge if the precision of the arithmetic used, represented by β^{-t}, is sufficiently high and that n, the condition number of A, and the growth factor g are not too large.

EXERCISES

E9.4.1   Define the 2×2 matrices

$$H_k = \begin{bmatrix} 0 & 2 \\ 0 & 0 \end{bmatrix}, \quad k \text{ odd}; \qquad H_k = \begin{bmatrix} 0 & 0 \\ 2 & 0 \end{bmatrix}, \quad k \text{ even}.$$

Show that the iterates (5), with $\mathbf{d}^k \equiv 0$, do not converge to $\mathbf{x}^* = \mathbf{0}$ for all $\mathbf{x}^0$ even though $\rho(H_k) = 0$, $k = 0, 1, \ldots .$

E9.4.2   Consider the system

$$0.832x_1 + 0.448x_2 = 1.00$$
$$0.784x_1 + 0.421x_2 = 0.$$

The exact solution, to three figures, is

$$x_1 = -439., \qquad x_2 = 817.$$

Solve the system using three-digit decimal unrounded arithmetic; your computed solution should be

$$x_1^0 = -506., \qquad x_2^0 = 942.$$

Now carry out two steps of iterative refinement, computing the residuals in double precision and chopping back to single precision. Show that

$$r_1^0 = -0.024, \qquad r_2^0 = 0.122, \qquad x_1^1 = -429, \qquad x_2^1 = 798.$$
$$r_1^1 = 0.424, \qquad r_2^1 = 0.378, \qquad x_1^2 = -439, \qquad x_2^2 = 819.$$

Note that one cannot detect the improvement via the residuals.

READING

The basic references for rounding error analysis for gaussian elimination are Wilkinson [1963] and Wilkinson [1965], which also include analyses of methods for computing roots of polynomials and eigenvalues of matrices. See also Forsythe and Moler [1967] for a very readable summary of Wilkinson's work on linear equations. For a theoretical development of gaussian elimination and related direct methods see Householder [1964].

BIBLIOGRAPHY

Coddington, E., and Levinson, N. (1955). "Theory of Ordinary Differential Equations," McGraw-Hill, New York.

Daniel, J. (1971). "The Approximate Minimization of Functionals," Prentice-Hall, Englewood Cliffs, New Jersey.

Dieudonné, J. (1969). "Foundations of Modern Analysis" (Enlarged and Corrected, Printing), Academic Press, New York.

Faddeev, D., and Faddeeva, V. (1960). "Computational Methods of Linear Algebra," Fizmatgiz, Moscow. Translated by R. Williams, Freeman, San Francisco, California, 1963.

Forsythe, G., and Moler, C. (1967). "Computer Solution of Linear Algebraic Systems," Prentice-Hall, Englewood Cliffs, New Jersey.

Forsythe, G., and Wasow, W. (1960). "Finite Difference Methods for Partial Differential Equations," Wiley, New York.

Gantmacher, F. (1953). "The Theory of Matrices," Gosud. Izdat. Tehn.-Teor. Lit., Moscow. Translated by K. Hirsch, Chelsea, New York, 1959.

Gear, C. (1971). "Numerical Initial Value Problems in Ordinary Differential Equations," Prentice-Hall, Englewood Cliffs, New Jersey.

Hahn, W. (1967). "Stability of Motion," Springer-Verlag, Berlin and New York.

Henrici, P. (1962). "Discrete Variable Methods for Ordinary Differential Equations," Wiley, New York.

Householder, A. (1964). " The Theory of Matrices in Numerical Analysis," Ginn (Blaisdell), Boston, Massachusetts.

Isaacson, E., and Keller, H. (1966). "Analysis of Numerical Methods," Wiley, New York.

Keller, H. (1968). "Numerical Methods for Two-Point Boundary Value Problems," Ginn (Blaisdell), Boston, Massachusetts.

Ortega, J., and Rheinboldt, W. (1970). "Iterative Solution of Nonlinear Equations in Several Variables," Academic Press, New York.

Varga, R. (1962). "Matrix Iterative Analysis," Prentice-Hall, Englewood Cliffs, New Jersey.

Wendroff, B. (1966). "Theoretical Numerical Analysis," Academic Press, New York.

Wilkinson, J. (1963). "Rounding Errors in Algebraic Processes," Prentice-Hall, Englewood Cliffs, New Jersey.

Wilkinson, J. (1965). "The Algebraic Eigenvalue Problem," Oxford Univ. Press (Clarendon), London and New York.

Young, D. (1971). " Iterative Solution of Large Linear Systems," Academic Press, New York.

INDEX

Page numbers given in italics indicate the pages on which the appropriate definitions may be found.

197